Front Panel

Designing Software for Embedded User Interfaces

Niall Desmond Murphy

R&D Books
Lawrence, Kansas 66046

R&D Books
an imprint of Miller Freeman, Inc.
1601 West 23rd Street, Suite 200
Lawrence, KS 66046
USA

Designations used by companies to distinguish their products are often claimed as trademarks. In all instances where R&D is aware of a trademark claim, the product name appears in initial capital letters, in all capital letters, or in accordance with the vendor's capitalization preference. Readers should contact the appropriate companies for more complete information on trademarks and trademark registrations. All trademarks and registered trademarks in this book are the property of their respective holders.

Copyright © 1998 by Miller Freeman, Inc., except where noted otherwise. Published by R&D Books, an imprint of Miller Freeman, Inc. All rights reserved. Printed in the United States of America. No part of this publication may be reproduced or distributed in any form or by any means, or stored in a database or retrieval system, without the prior written permission of the publisher; with the exception that the program listings may be entered, stored, and executed in a computer system, but they may not be reproduced for publication.

The programs in this book are presented for instructional value. The programs have been carefully tested, but are not guaranteed for any particular purpose. The publisher does not offer any warranties and does not guarantee the accuracy, adequacy, or completeness of any information herein and is not responsible for any errors or omissions. The publisher assumes no liability for damages resulting from the use of the information in this book or for any infringement of the intellectual property rights of third parties that would result from the use of this information.

Cover art created by Robert Ward.

Distributed in the U.S. and Canada by:
Publishers Group West
P.O. Box 8843
Emeryville, CA 94662
ISBN: 0-87930-528-2

un Miller Freeman
A United News & Media company

Table of Contents

Chapter 1   *Introduction* *1*
                 1.1 The Audience 2
                 1.2 Software Complexity 2
                 1.3 The Programmer and Usability....................... 3
                 1.4 The Desktop Environment versus the
                 Embedded Environment................................ 3
                 1.5 Code Examples 4
                 1.6 The Text 5
                 1.7 Real-Time Operating Systems 5
                 1.8 Feedback... 5
                 1.9 Acknowledgments.................................... 6

Chapter 2   *The Model-View-Controller Paradigm* *7*
                 2.1 What Does Object-Oriented Software Mean
                 in a User Interface? 7
                      2.1.1 Encapsulation 9
                      2.1.2 Polymorphism............................. 10
                      2.1.3 Object-Oriented User Interfaces versus
                      Object-Oriented Implementations...................... 12
                      2.1.4 Recognizing Object-Oriented Design............ 12
                 2.2 Model-View-Controller 12
                      2.2.1 A Layered Description of the System............ 13
                      2.2.2 Identifying the MVC Objects 15
                 2.3 An Array of LEDs................................. 16
                      2.3.1 Optimizations of the Refresh Function............ 18
                 2.4 Layered Messages on a Small Text Display............. 18
                      2.4.1 The Text Display Manager Module 20
                 2.5 Conclusion 25

iii

Chapter 3 **Scheduling and Managing User Events 27**
　3.1　User Events on the Desktop 27
　3.2　Polling versus Interrupts. 28
　　　　3.2.1　Polling from the Main Line 28
　　　　3.2.2　Timed Polling. 29
　　　　3.2.3　Interrupts 30
　3.3　Queuing 32
　3.4　Is an RTOS Needed? 34
　3.5　The Tasks 36
　　　　3.5.1　Priority in a Tasking System 39
　　　　3.5.2　Allowing Multiple Tasks Access to the Display 39
　　　　3.5.3　Breaking Up Long Events into Shorter Events 40
　　　　3.5.4　A Task with a View 41
　3.6　Different Queues for Different Events. 43
　3.7　Queue Read-Ahead 44
　3.8　Directing Event Traffic 45
　　　　3.8.1　The Focus. 47
　　　　3.8.2　Callbacks 49

Chapter 4 **Finite-State Machines and**
　　　　　　　Table-Driven Software. 55
　4.1　The FSM as a Poor Man's Real-Time Operating System . . . 55
　4.2　Drawing the FSM. 56
　4.3　Why User Interface Code Needs FSMs 57
　　　　4.3.1　States and Modes in the User Model 58
　4.4　Limits of the FSM 58
　4.5　How Many States? How Many FSMs? 59
　4.6　Use of Constant, Static, Global, and Automatic Data 60
　4.7　Example of an FSM for a Toggle Button on a GUI. 60
　　　　4.7.1　The Four States: Up, Down, Pending_Up, and
　　　　　　　Pending_Down 61
　　　　4.7.2　The Interaction Between the FSMs 62
　　　　4.7.3　The Implementation. 63
　　　　4.7.4　A Model-View-Controller Interpretation
　　　　　　　of this FSM. 66
　4.8　Menu for a Small Text Display 66
　　　　4.8.1　Human Factor Issues for Menus 67
　　　　4.8.2　The Implementation. 68
　　　　　　　4.8.2.1　The Model 68
　　　　　　　4.8.2.2　The View 76
　　　　　　　4.8.2.3　The Controller 78
　　　　4.8.3　Advantages of Table-Driven Code. 79

Chapter 5 **Graphics** *81*
        5.1 The Software Levels 82
        5.2 Choosing the Set of Primitives 82
                5.2.1 What Do I Want to Draw? 83
                5.2.2 Support for Flexible Drawing 84
        5.3 The Next Level Up — Do You Need Objects?. 85
                5.3.1 Structures to Define Graphic Objects 86
                5.3.2 Memory Management and Initialization. 90
                5.3.3 Container Hierarchies 93
        5.4 Refreshing the Display 95
                5.4.1 Refreshing by Object 96
                5.4.2 Refreshing by Area 105
                5.4.3 When to Refresh? 106
        5.5 Compound Objects 106
                5.5.1 Button 106
                        5.5.1.1 Locating Events 108
                5.5.2 Line Graphs. 116
        5.6 Touch Screen Programming Techniques 121
        5.7 Multiple Dialogs 124
        5.8 Conclusion 127

Chapter 6 **Systems Issues** *129*
        6.1 Multiple Processors. 129
                6.1.1 Protocols 131
                        6.1.1.1 Text Protocols. 132
                        6.1.1.2 Stateless Protocols 134
                6.1.2 Multiple User Interfaces 135
        6.2 Safety 136
                6.2.1 Safety versus Availability versus Reliability 137
                6.2.2 Sanity Checks 138
                6.2.3 Fail-Safe State. 140
                6.2.4 Hazard Analysis 141
                        6.2.4.1 Fault Tree Analysis (FTA) 141
                        6.2.4.2 Failure Mode Effects Analysis (FMEA). 143
                6.2.5 Interlocks. 143
                6.2.6 Watchdogs. 144
                6.2.7 Roles and Security 146
                6.2.8 Separate Channels of Control and Monitoring 146
                6.2.9 The Complexity Paradox 150

	6.3 Translation and Internationalization...................151
	6.3.1 Storing the Strings 152
	6.3.2 Character Sets 153
	6.3.3 Localizing Hardware 154
Chapter 7	***C++ for Embedded Systems*** *157*
	7.1 Introduction to C++ 158
	7.1.1 Classes 158
	7.1.1.1 Constructors and Destructors.............162
	7.1.2 Inheritance 165
	7.1.3 Virtual Functions 168
	7.1.4 Templates.....................................170
	7.1.5 Exceptions 172
	7.1.6 Other Features 175
	7.2 Choosing C++ 176
	7.2.1 C++ for User Interfaces 176
	7.2.2 Programming in the Large 177
	7.2.3 The Performance Question:
	Does C++ Really Lead to Bigger, Slower Executables?...178
	7.2.3.1 In-Line Functions.......................179
	7.2.3.2 Code Reuse Effects on Efficiency 180
	7.2.3.3 Memory Management 181
	7.2.3.3.1 Operator new 181
	7.2.3.3.2 Heaps and Fragmentation 184
	7.2.3.3.3 Tasking 185
	7.2.4 Code Reuse and Complexity 186
	7.2.4.1 Multiple Inheritance.....................186
	7.2.4.2 Templates..............................186
	7.2.4.3 Exceptions 187
	7.2.4.4 Taming the Complexity 188
	7.3 Using C++ 189
	7.3.1 Implementing a Focus, or Callbacks, in C++.......189
	7.3.1.1 C Callbacks from C++ 192
	7.3.2 Hardware Issues...............................193
	7.3.2.1 Start-up................................193
	7.3.2.2 Encapsulation...........................193
	7.3.2.3 Classes in ROM........................194
	7.3.2.4 Placement new Operator..................195

	7.4 Converted Examples.............................. 197	
	7.4.1 Text Display Module 198	
	7.4.2 `multi` Graphics Demo 202	
	7.4.2.1 The `Display` Class..................... 209	
	7.4.2.2 The `Container` Class 209	
	7.4.2.3 Inheriting from `Container`.............. 211	
	7.4.2.4 Lists of Objects....................... 213	
	7.5 Conclusion 215	
Chapter 8	***The Design Process*** *217*	
	8.1 Writing Requirements............................. 220	
	8.1.1 Walkthroughs 221	
	8.1.2 Testability................................ 221	
	8.1.3 Assumptions About the User 222	
	8.1.4 Learnability............................... 223	
	8.1.5 Maintaining a Glossary...................... 224	
	8.2 Iterations Through the Interface 224	
	8.2.1 Test Team's Role 226	
	8.3 Simulations and Prototypes......................... 227	
	8.3.1 Simulations............................... 227	
	8.3.2 Prototypes................................ 229	
Chapter 9	***Usability for Embedded Systems*** *231*	
	9.1 The User 234	
	9.1.1 How Keen is Your User? 234	
	9.1.2 Roles.................................... 236	
	9.1.2.1 Novice and Expert 236	
	9.1.2.2 Are You Being Served? 237	
	9.1.2.3 Privileged Users 238	
	9.1.3 The User's Conceptual Model 238	
	9.1.4 Observing Users 240	
	9.1.4.1 Users, Users Everywhere 240	
	9.1.4.2 Getting Feedback 241	
	9.1.4.3 Paper Prototypes and Working Prototypes... 242	
	9.1.4.4 Interpreting the Feedback................ 243	
	9.1.4.5 Usefulness and Usability 244	

9.2 The Interface 244
    9.2.1 Robustness 245
    9.2.2 Consistency 247
    9.2.3 Affordance 247
    9.2.4 Icons 248
    9.2.5 Surface Area 250
    9.2.6 Compatibility 251
    9.2.7 Setting the Pace 253
    9.2.8 Paths Through the Interface 254
        9.2.8.1 Directed Interfaces 254
        9.2.8.2 Multithreading 254
        9.2.8.3 Modes 256
        9.2.8.4 Equal Opportunity 256
        9.2.8.5 Multiple Paths 257
    9.2.9 Migrating from Mechanical Controls 257
9.3 The Graphical User Interface 258
    9.3.1 Disadvantages of the GUI 259
    9.3.2 Getting the Most Out of a GUI 260
        9.3.2.1 Color 263
    9.3.3 Windows 263
    9.3.4 Pointerless 265
    9.3.5 Touch Screens 267
9.4 Usability and Safety 268
    9.4.1 The Usability versus Safety Trade-off 269
    9.4.2 Validating Input 270
    9.4.3 Monitoring 272
    9.4.4 Alarms 274
        9.4.4.1 Get Your Priorities Right 274
        9.4.4.2 Text Lists versus Dedicated Annunciators ... 275
        9.4.4.3 Cry Wolf 276
        9.4.4.4 Testing Alarms 276
    9.4.5 Discovering Safety-Critical Mistakes 278

Chapter 10   *The Future* *281*
10.1 Working Together 282
10.2 Graphics Everywhere 283
10.3 Voice Synthesis and Verbal Commands 284
10.4 Nuts and Bolts 285
10.5 Software Quality 286
10.6 Programming Activities 287

Appendix A   *RTOSs and μC/OS* **289**
        A.1 Real-Time Operating System Concepts 289
            A.1.1 Queues 290
            A.1.2 Priority 290
            A.1.3 Reentrancy 291
        A.2 μC/OS 291
            A.2.1 Start-Up 291
            A.2.2 `OSInit()` 293
            A.2.3 `OSQCreate()` 293
            A.2.4 `OSQPend()` 294
            A.2.5 `OSQPost()` 294
            A.2.6 `OSStart()` 294
            A.2.7 `OSTaskCreate()` 295
            A.2.8 `OSTimeDly()` 295

Appendix B   *The Companion Disk* **297**
        B.1 Installation 297
        B.2 The Directory Structure 298
            B.2.1 `chap2/tdm` 299
            B.2.2 `chap3/keyq` 299
            B.2.3 `chap3/watch` 300
            B.2.4 `chap4/fsm` 300
            B.2.5 `chap4/menu` 300
            B.2.6 `chap5/multi` 301
            B.2.7 `chap7/multi` 301
            B.2.8 `chap7/tdm` 302
            B.2.9 `common` 302
            B.2.10 `include` 302
            B.2.11 `ucos` 302
            B.2.12 `borland3` 302
            B.2.13 `borland4` 302
        B.3 Licensing 303

Bibliography *305*

Glossary *309*

Index *311*

Chapter 1

Introduction

Welcome to *Front Panel: Designing Software for Embedded User Interfaces*. This book is for programmers and engineers working on embedded systems. The front panel is the part of the device (e.g., VCR, photocopier, or digital watch) the user looks at and interacts with. It is where you put the flashing lights, the knobs, and the dials, because this is where they can be seen, not on the back panel, where the untidy cable connectors live. Desktop computers have a standard screen, keyboard, and mouse to allow user interaction. In the embedded world, there are many interfaces, from the single button remote control for a car alarm to screens of information and banks of switches in industrial and power plant control rooms. Applying one name to the array of user interfaces that fall between is impossible, but often those controls sit on a single panel, and that is where the user meets the machine. For this book, a front panel is any physical part of the system that enables the user to control or observe the device via software.

This book is primarily about designing software; however, building embedded systems is a multidisciplinary activity, requiring mechanical, electronic, and software skills. I occasionally step outside the brief of this book to discuss hardware issues, but only those that might influence the requirements or implementation of the software.

User interface design, distinct from software design, applies to the appearance of the front panel and to the interactions that occur between the user and the device. The visual appeal of the front panel is not addressed in this book because it does not affect the software in any way. The interactions, however, do have a big impact on software. They decide how software must react to certain buttons or dials. Chapter 9, "Usability for Embedded Systems," investigates that area in depth, but I make occasional references to the usability of features as they are discussed in the rest of the book. It is not always possible to separate the concerns of software design from the requirements that drive that software.

1.1 The Audience

I have encountered two types of programmer in the embedded world. One is the electronics designer, who has to write firmware to drive the electronics he has designed. Gradually, the amount of software design grows, the amount of electronics design shrinks, and the electronics engineer has turned into a software engineer. These guys are vital on a software team because you need someone who knows the system electronics inside out.

The second type of engineer is the desktop programmer with a computer science background who has changed jobs to write software for a device that is no longer a computer, but a device dedicated to a single job. These engineers are often shocked by the lack of facilities that are available to them. This second type of engineer is also valuable to the project. That is because the desktop world has already gone through much of the evolution that the embedded software world is now experiencing. The lessons of code structure, version control, and managing programs that run to tens of thousands of lines of code are more quickly learned on the desktop than on embedded systems. This is partly because a desktop application contains far more features than a typical embedded system. Also, the stability of the desktop development environment allows larger programs to be written in less time.

You can probably identify yourself in one of these two roles, and most of your colleagues will fall into one of the camps also. I hope this book has something to offer both. The descriptions of object orientation and discussions of how to manage event-handling code may involve more structure than was required in low-level firmware. The performance discussions will help the ex-desktop programmer realize that he can no longer allocate megabytes of RAM for a task without serious consideration. He is now in a world in which the software is given away as part of a physical device and in which the cost of parts, including memory, has a serious impact on the marketability of the device.

1.2 Software Complexity

Apart from user interface software, most of the software written for embedded systems is for process control. The fundamental difference between the two is that user interface code is event-driven, while process control is more procedural. This may sound like a gross generalization because many processes have to observe alarms, or interrupts, from external signals. But the meanings of those alarms and interrupts rarely change. On the user interface, the meaning of every action depends on the context that was created by the sequence of events that went before. Each event changes the pattern of the front panel in some large or small way, like moving a piece on a chessboard. And like the chess game, the number of possible sequences of inputs and resulting states from an apparently simple front panel quickly multiplies to a number that cannot be investigated individually. Although hardware complexity is limited by the budget allocated to the components that make up the front panel, no such natural limit applies to software.

This complexity puts a strain on the design of the interaction and on the design of the software. This book provides you with some of the tools required to manage that complexity.

1.3 The Programmer and Usability

While this book addresses the software techniques that can be employed in developing an interface, I also hope to encourage the reader to think big. There is little benefit in putting a lot of time into making a keypad-reading routine as efficient as possible only to find out later that the customer actually wants a pointing device. Ideally, these two functions are separate parts of the design process and can be performed by different people. In practice, many organizations do not hire usability professionals. Even when much of the interface is guided by a nonprogrammer, it is ultimately the programmer who must weigh the amount of effort and resources it will take to implement an interaction. If the programmer knows nothing about usability, he cannot communicate effectively with the usability representative. At the other extreme, all the interactions are designed completely by the programmer. When a graphics screen is available, there is almost no restriction put on the programmer by the electromechanical or industrial engineer who guided the front panel layout of buttons, dials, and lights. As user interfaces become more sophisticated, programmers must fully understand usability and the target user.

1.4 The Desktop Environment versus the Embedded Environment

MS-Windows and X-Window provide much of the functionality required to build the user interface, allowing the desktop programmer to concentrate on the application-specific portion of the program. Embedded environments cannot exercise the same amount of code reuse. The hardware environment is often significantly different from any desktop environment. Even if code could be ported to the target processor, the hardware used for input and output is so radically different that the GUI toolkits would be rendered useless except in some cases in which the embedded system uses a standard graphics controller.

This book seeks to provide the information needed to tackle user interface problems you would be protected from if you were working on the desktop. Mechanisms for deciding when to refresh the display and for queuing events are addressed in Chapters 2 and 3. Usability issues are also different for embedded systems, because the physical devices the user interacts with are more diverse than the standard keyboard, screen, and mouse.

1.5 Code Examples

Chapters 2 through 5 assume that the reader is familiar with C programming. However, the algorithms and ideas are, for the most part, applicable to other procedural languages and assemblers.

Larger examples are supplied on the companion disk. Appendix B describes the directory structure and how to install the software. I have avoided reproducing the listings in full in the book, as large listings might discourage you from reading further. Instead, I have interspersed portions of the code with the descriptions at the appropriate point. Code listings on the disk are lengthier than the sections presented in the book because I wanted to present examples that were part of a real interaction. Although many assumptions and simplifications are made in order to keep the size of the programs reasonable, I felt that if I did not make the example code part of an executable program, there was a danger of delivering code that was too abstract for the reader to see its applicability. The problem with producing an executable program is that there is a certain amount of dull code that has to be written to fulfill some less crucial goals. I try to avoid boring you with that in the book. They are on the disk for your perusal if you are curious.

The book has been written to be self-contained and complete. I do indicate a number of places at which it may be helpful to run a demo on the disk. I realize that not everyone has access to the PC platform required to run the programs. In those cases, the summaries of each program in Appendix B will be helpful.

I have applied a few coding conventions to my examples to make them more readable. I do not necessarily advocate these standards for your projects, although you should always apply some naming convention, however simple. I use the prefix `G_` on global variables. Static variables that have file scope are preceded by `GS_`. I append `Ptr` to the names of pointers variables. I would not recommend this rule for your production code, but it helps emphasize the pointers in example code. In Chapter 7, "C++ for Embedded Systems," I append a single underscore to private data members.

I use the `assert()` macro regularly in the code examples to assist debugging and to clarify the design assumptions of the piece of code.

```
assert(x < 5);
```

This line means that the design of the code assumes that `x` is always greater than 5. For more information on the `assert()` macro, see Section 6.2.2 "Sanity Checks" on page 138.

Occasionally, a single C statement runs to more than one line on the page. This inhibits readability, but it is the result of a compromise. Less space would have been taken up if I had reduced the font size or shortened variable names. I felt that either measure would have been more detrimental to readability than wrapping the lines.

1.6 The Text

Within the text, I have emphasized all type and variable names in order to disassociate these names from any meaning they may have in normal English. My apologies for the jarring appearance of some sentences caused by this convention.

In more academic writing, the use of the pronoun he or she is avoided. The frequency with which I have to refer to the actions of users makes it difficult to write without one of these pronouns. Although I considered using she throughout or mixing the two, I decided to go for the more conservative approach. On many occasions, I refer to mistakes and user errors. An implication that female users are more likely to make mistakes when using technologically sophisticated devices reflects neither my opinions nor the real world. Please read he to mean a person of either gender.

1.7 Real-Time Operating Systems

Some of this book, particularly Chapter 3, assumes a rudimentary knowledge of real-time operating systems. If you want more information on real-time operating systems or on µC/OS, the real-time operating system used in my examples, turn to Appendix A.

1.8 Feedback

If you find mistakes in the code or in the text, or if you disagree with some of my ideas, feel free to contact me at `nmurphy@iol.ie`. I have a Web page at `http://www.iol.ie/~rmurphy`. I am also interested in hearing from anyone who has put the ideas in this book to use in an interesting or novel way.

1.9 Acknowledgments

The following people contributed to this book, either directly or indirectly, in the opportunities and education they provided to me: Gerry Fahy, Patrick Murray, John O'Dea, David Miller, Eddie Carroll, Paul Bennett, Steven Huggins, and Thad Smith. Thanks to R&D Books, Berney Williams, and Liza Niav for taking a collection of ideas, word-processed documents, and code fragments and turning them into a book.

I would like to thank Jean Labrosse for the use of µC/OS and for material from his book, *µC/OS: The Real Time Kernel*. I would also like to thank Tyler Sperry and Lindsey Vereen who, as editors of *Embedded Systems Programming* magazine, provided me with the opportunity to publish some of the ideas that I explore further in this book.

Thank you, Mum and Dad, for opening all doors.

Chapter 2

The Model-View-Controller Paradigm

This chapter lays the foundation for the two most important design techniques used in user interface software. I introduce object-oriented design and then the model-view-controller paradigm, which is used in many of the following chapters. This paradigm is the key to decomposing user interface design into several subsystems with clean interfaces.

2.1 What Does Object-Oriented Software Mean in a User Interface?

The adjective *object-oriented* is overused and has been rendered almost meaningless; marketing departments label almost any software "object-oriented." In terms of the problems that are addressed in this chapter and in the rest of the book, "object-oriented" really means that the program's data structures contain fields that describe the things the user sees or manipulates.

This example should solidify the previous definition. If an interface has eight LEDs, they may be represented by eight values. The values are on and off. Implemented in assembler, each bit represents one LED. It could also be implemented in C++, with a class of type LED. Eight instances are created, memory is allocated, constructors are called, and member functions are used to get the current state of the LED

7

and alter it. Both implementations are object-oriented because the design matches the items or objects that the user can perceive. You could conceivably point to a piece of memory, possibly a single bit, and say, "That is the 'motor running' LED."

The assembler implementation has the advantage of efficiency. The C++ implementation is easily expanded to capture other concepts, such as flashing, multicolored LEDs and a myriad of other features that may be added to the device at a later date. It also reduces the risk of accidentally turning off all the other LEDs when turning on the "motor running" LED.

An alternative implementation is to maintain a list of numbers, as seen in Figure 2.1. The numbers in the list refer to LEDs that are on. This may be an efficient implementation. If it is common to have no LEDs illuminated, a single check can tell that none are illuminated and no further checking is required. There is no need to loop through each LED to check its status. However, no matter how appropriate this implementation is, it is not object-oriented.

The motor running LED no longer has a location in memory devoted to it. This does not mean it is a bad design or that it should not be used. That final judgment lies with the designer. Purists argue that the actual representation in memory is irrelevant. However, mapping the visible artifacts on the user interface to memory in this way is a good first step to understanding how an object-oriented implementation differs from a non-object-oriented interface.

Although the ability to identify a piece of memory that represents the entity helps you to recognize an object-oriented design, it does not completely define the objects themselves. An object consists of code as well as data. The code is related to the data

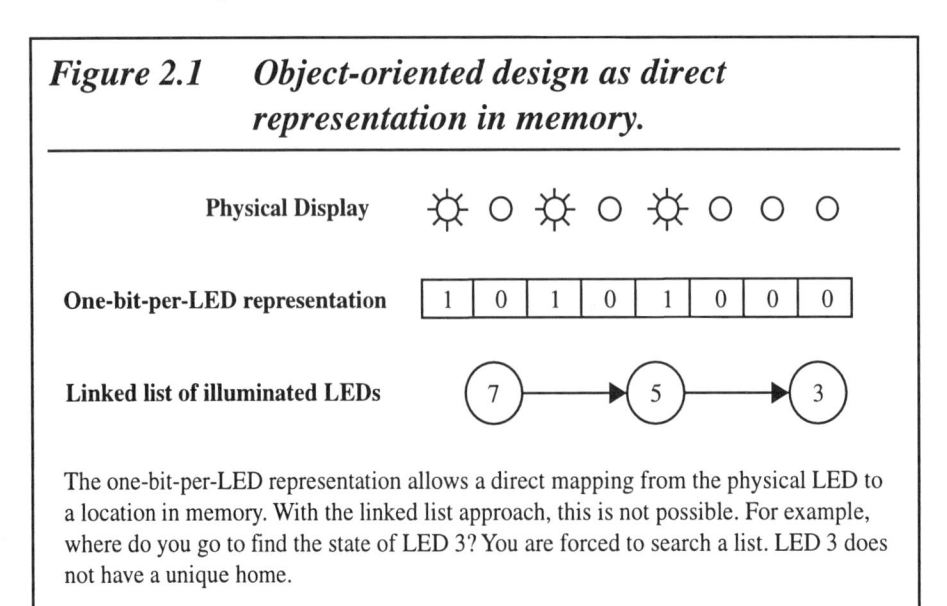

Figure 2.1   Object-oriented design as direct representation in memory.

Physical Display	☼ ○ ☼ ○ ☼ ○ ○ ○
One-bit-per-LED representation	1 0 1 0 1 0 0 0
Linked list of illuminated LEDs	7 → 5 → 3

The one-bit-per-LED representation allows a direct mapping from the physical LED to a location in memory. With the linked list approach, this is not possible. For example, where do you go to find the state of LED 3? You are forced to search a list. LED 3 does not have a unique home.

because the code manipulates the data. In object-oriented languages, it is possible to define which code belongs to which objects and, therefore, which pieces of data the code can access. In C, you can make your code more object-oriented by keeping all the code that manipulates a data structure in one "place." The place may mean file, library, or directory, depending on the level at which you work.

I hope my example illustrates that it is not necessary to have a sophisticated language, such as C++, or a completely object-oriented environment, such as Smalltalk, to exploit object-oriented techniques in your code.

The more traditional way to describe object-oriented programming is in terms of encapsulation, polymorphism, and inheritance. Inheritance is not a concern until Chapter 7, so I will defer its definition until then.

2.1.1 Encapsulation

Encapsulation means that the part of the program concerned with lighting LEDs protects other parts of the program from having to know anything about lighting LEDs. The interface to the LED-controlling module should not depend on the actual implementation. Most programmers simply call this common sense and have been doing it for years without giving it a name as grandiose as encapsulation.

The implementation of the LEDs' controlling module includes the data structure that stores the state and the related functions that operate on that data. If there are eight LEDs, there are eight objects. Their data may be stored in the same format, but there is a distinct piece of data for each object.

Providing encapsulation often means providing a set of functions that provides a service and has an external interface that does not necessarily expose the internal implementation. A disadvantage of this, if applied at too fine a level, is that many more function calls are required. Boundaries are put in place to provide protection, but now a small price is paid each time the flow of control passes a threshold. If optimization is a concern, some of this overhead can be avoided by providing macros rather than functions. Macros have their own disadvantages because the compiler does not provide type checking for macro arguments. A bigger optimization issue is that if the interfaces to a set of functions do not provide a good match for actual use, the programmer may miss the opportunity to take some shortcuts. Consider encapsulating a set of four Boolean variables, each stored in a single bit. The calling function may regularly set all four Booleans together. If the available function only allows the Booleans to be set individually, the four operations will require four separate calls to that function to mask in the new value. Had the Booleans not been protected by a function call interface, a more optimal approach could have been taken. What encapsulation does provide in this case is the ability to add some action to be performed any time one of the Boolean changes value. The calling modules do not have to be aware that this action was added.

C provides only a few constructs that support encapsulation. Variables or functions that are declared `static` within the file scope cannot be accessed outside that file. Variables that are declared `static` within a function cannot be accessed from outside the function, but they hold their values between calls. Even though programmers can choose to honor the interfaces set up to manipulate certain structures, it is better to catch any violations at compile time because they may be difficult to spot in code inspections and impossible to detect from the behavior of the running system. Object-oriented languages such as C++ provide more features for protecting data at various levels, with the guarantee that violations will be flagged during compilation.

2.1.2 Polymorphism

Polymorphism is the property of acting like one type at one time and another type at another time. More importantly for software, polymorphism implies the ability to treat a number of slightly different things as if they were the same.

Assume that you have a system that has several LED indicators and several incandescent bulbs, as seen in Figure 2.2. Assume that the LEDs are multiplexed and controlled differently from the bulbs. Each has their own type of data structure. You then write a single device driver capable of controlling either type of light. You could now provide the rest of the system with an interface that consists of `setLEDOn()`, `setLEDOff()`, `setBulbOn()`, `setBulbOff()`. The caller would have a way of indexing the LEDs and bulbs. For example, there are three LEDs, numbered 1, 2, and 3, and three bulbs, numbered 1, 2, and 3. I will call this design A.

Figure 2.2   Polymorphism: Are LEDs and bulbs the same or different?

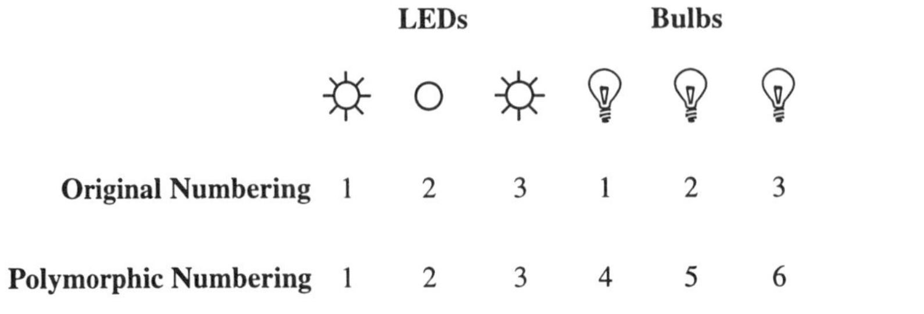

	LEDs			Bulbs		
Original Numbering	1	2	3	1	2	3
Polymorphic Numbering	1	2	3	4	5	6

The two types of lights can be indexed as two sets of three lights, or as one set of six lights.

Instead of dealing with LEDs and bulbs, it may be preferable to allow the caller to deal with lights, without distinguishing the type. You would supply the user with only two functions — `setLightOn()` and `setLightOff()`. The caller would have a method of indexing that did not distinguish between the two types of light. Now you have lights numbered 1, 2, 3, 4, 5, and 6. Any of them could be a LED or a bulb. The actual implementation could have a data structure per light that contains a union and a discriminating field that tells us which type of light it is. Alternatively, the index of the light tells us which type it is — for example, the first three are LEDs and the next three are bulbs. I will call this design B.

Now that I have defined a polymorphic light, what use is it? The benefit arises when writing code that applies an algorithm to both types of object, without caring about the distinction. Perhaps you wanted to write a routine that could flash a light on for one second. Using design A, you would have to know which type of light you are dealing with.

```
void flashLight(Boolean isLED, int index)
{
    if (isLED)
    {
        setLEDon(index);
    }
    else
    {
        /* not a LED so it must be a bulb */
        setBulbOn(index);
    }
    delay(ONE_SECOND);
    if (isLED)
    {
        setLEDoff(index);
    }
    else
    {
        /* not a LED so it must be a bulb */
        setBulbOff(index);
    }
}
```

For the polymorphic light in design B, you do not have to care what type it is.

```
void flashLight(int index)
{
    setLightOn(index);
    delay(ONE_SECOND);
    setLightOff(index);
}
```

The design B version of `flashLight()` can be used without knowing what kind of light it is. If a third type of light were introduced, this version of `flashLight()` would not have to be updated, while the design A version of `flashLight()` would have to allow for the third type with another `else` branch. If you need to know which type of light is being indexed, the device driver in design B could easily provide an `isLed(int index)` and an `isBulb(int index)` function that would return Booleans.

From this example, it is easy to see that interfaces can be polymorphic without any direct support from the language. Sometimes, it is of great advantage to treat many objects in the same way. Polymorphism contributes to code reuse, which is one of the most important goals of object-oriented software.

2.1.3 Object-Oriented User Interfaces versus Object-Oriented Implementations

The term object-oriented is occasionally applied to user interfaces as distinct from the implementation in software of those interfaces. It is prudent to distinguish the two because the object-oriented term is so overused. In some contexts, an object-oriented user interface refers to a user interface that is based on objects from the real world, such as calendars, clocks, or pieces of paper. In other contexts, the term defines an interface in which different parts of that interface can be independently manipulated. Neither of these definitions corresponds to the concepts of object orientation in software design or implementation. Constantine (1995) is as confused as I am. In this book, the term object-oriented will be applied to designs and implementations and to languages that support those designs, but never to the user interface itself.

2.1.4 Recognizing Object-Oriented Design

In the examples that follow, it is useful, though not necessary, to recognize where object-oriented principles have been put to use. Because C is not an object-oriented language, the boundaries of which piece of code belongs to one object and which belongs to another may be vague. However, the interpretation of the data structures should be unambiguous.

Hopefully, this section has given you an introduction to the meaning of *object-oriented*. You will not need a more detailed understanding until Chapter 7, in which I deal with the support for object-oriented programming in C++.

2.2 Model-View-Controller

A concept I will explore in more detail is the model-view-controller (MVC) paradigm. The term originated in the Smalltalk community, but the concepts are applicable in any programming language. First, I provide a full description of the MVC paradigm, then I provide a few examples that show how it may map onto software for an embedded system. The MVC paradigm was originally intended for graphical interfaces, but you shall see that the principles are no different when applied to custom displays.

The model is the object that describes the information that the user may access or manipulate. Those data structures comprise all the information accessible from the display. The view is the object that represents what the user actually sees. The view can be thought of as a filter on the data in the model. The controller receives and manages the user's input. These inputs may require the controller to request changes to the data in the model or to the appearance of the view. For example, the model contains sales figures for ice cream broken down by flavor over the past 12 months. The view maps this data to a pie chart. The controller receives events from the user. The user is allowed to look at the data in a number of ways.

When a change is made to the model (for example, when this month's sales figure is updated), the view is informed and it is the view's responsibility to decide what changes, if any, must be made to the display. The change could have originated from user input via the controller or it could have come from another part of the application.

The controller may receive a mouse click that changes the pie chart from displaying sales by flavor to sales by month. Because the basic data stored in the model has not changed, the controller has to communicate with the view only to instruct it to update its display in the new format. However, when the user enters the latest sales figures for this month, those figures must be communicated to the model. In this case, the view must also be refreshed to reflect the new data in the model. Changes that originate with the user and are fed back to the user are U-turns. They can have varied depth. The physical feedback of a button moving when it is pressed is feedback that does not have to leave the mechanical domain. Other actions may be communicated to the application, where they will make permanent changes to the data being manipulated before the user sees the effect.

Although the model, view, and controller are all considered independent objects, the view and controller might not contain any state information or data members. They may simply perform a mapping between the physical layer and the model. In this example, the view would probably store the type of pie chart currently being rendered, while the controller stores no data.

2.2.1 A Layered Description of the System

Associated with the controller and the view are device drivers that allow the view to be displayed and allow the controller to get access to the devices manipulated by the user. This device driver level is also known as the physical layer. In a graphics program, the physical layer starts at the calls to `drawLine()` and `drawArc()`. On a custom display, the physical layer may start with calls to a function that manipulates the hardware, or the view may directly manipulate a bit that toggles the LED if the view and physical layer are merged.

For the controller, the physical input layer may perform filtering, such as key debouncing. The controller's job is not to control the hardware but to interpret the event's logical meaning and to decide how that event controls the model and/or the view. A model that is independent of the physical layer is good encapsulation and may make modifications to the code easier if the hardware changes. Having said that, picking a model that maps easily to the physical layer can make the code more efficient. If the states of a number of LEDs are controlled by a structure of 1 bit per LED, then storing them the same way in the model allows you to update eight LEDs in one operation. If the hardware uses 1 byte per LED, then copying that in the model may make the code more efficient because you do not have to use masking operations to examine individual bits. You have to make a judgment on the chances of the hardware configuration remaining the same and on the need for efficiency.

I will not go into much detail about the physical layer; Labrosse (1995) covers many common devices and programming techniques used in this area. The examples that follow in this chapter concentrate on the model and the view. The controller receives more attention in Chapter 3.

Above the level of the model is the application layer. This code need not be aware of the MVC triad's interactions with the outside world. However, it is common for some of the application-specific data to reside in the model. If a device has a speed setting that is displayed, the model may hold that value for use by the view, as well as for use by the rest of the application.

Figure 2.3 shows the model, view, and controller with arrows that show the flow of information or events. Note that the controller and view have no interaction with the application layer and that the model has no interaction with the physical layer. This is an example of the encapsulation just discussed. Events from the controller have to go to the application via the model because the model contains the context for the event. For example, a button might have different meanings at times. A pointer event on a graphics screen has to be resolved to an object in the model before it has any real meaning.

The lines between the model and view or between the model and the application are not always as clear-cut as in Figure 2.3. The overhead, in terms of code size, as well as run-time efficiency, is not always justified on small programs. For small systems, it is often useful to design in terms of the MVC paradigm but to code in a much tighter manner. In this way, you do not lose sight of the purposes of different parts of the system, but at run time you do not pay the price of enforcing all the boundaries. In some cases, the physical layer may be distinct, with a lot of the rest of the system merging. In other cases, the controller may be the most complex part and may need to be isolated from the rest of the system.

As programs get larger, the boundaries become more distinct. At the extreme, each subsystem runs on a separate processor, possibly running in separate units or locations. Consider a person who gains entry to a building by entering a personal identification number. A central security room several miles away may display part of the view of the model maintained in the security device. You may choose to look at this as one model with two views; one local, for feedback to the person entering the building, and one remote, for security monitoring.

2.2.2 Identifying the MVC Objects

This definition of the model-view-controller paradigm is loose compared to that you might learn on a Smalltalk system. Smalltalk dictates that the model, view, and controller each consist of its own object. In Smalltalk, objects have strict boundaries and cannot access each other's data. Typically, a system consists of several MVC triads. This is partly because it is a windowed system, and the graphical elements displayed in one area of the screen might bear little or no relationship with the elements in another area or with elements that are displayed at another time. Nongraphical embedded systems tend to be simpler, so you could consider it to have one model, one view, and one controller. In other cases, the front panel could be divided up so that

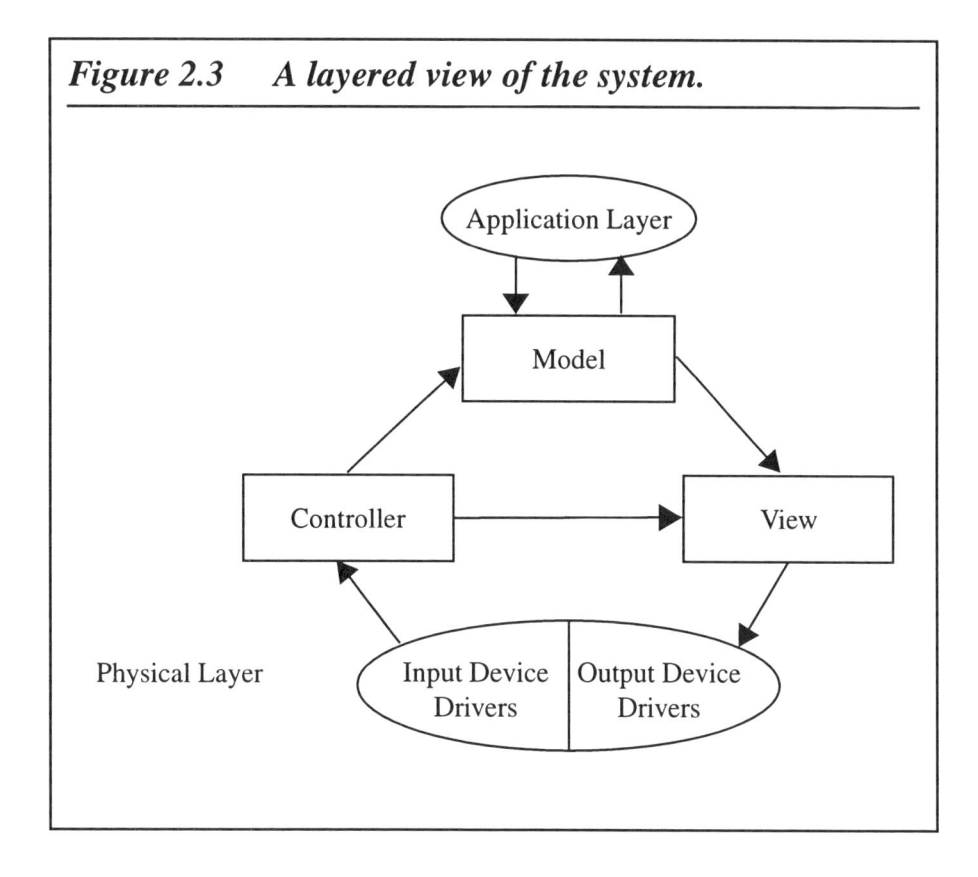

Figure 2.3   A layered view of the system.

you arrive at a system with several models represented by several views. In C, the difference between several structures that contain part of one big model and several structures that represent individual small models is more a question of semantics than of practicality. In Smalltalk and C++, object boundaries are more strictly defined, and the exact number of objects is less disputable.

Even when the boundaries have almost disappeared because your system is very small, has serious optimization constraints, or is written in assembler, it should still be possible to look at an operation and say: "That is updating the model," "That is changing the view based on an event in the controller," or "This array forms the model data, while that integer affects only the view and not the model itself." The ability to analyze the software in this way makes it easier to predict the impact of any code changes.

The following examples in this chapter will put these abstract ideas into practice.

2.3 An Array of LEDs

Now that the user interface has been broken down into its most basic parts, I will try to identify some of them in a simple example. Consider a display in which a set of LEDs is under software control. Each LED represents some Boolean condition within the system. The LEDs may be on or off, reflecting a true or false value of the Boolean. To represent this information in the model, I can use a simple array of Booleans. If an enumerated type gives a name to each Boolean, that enumerated type can be used to access the array. For example, I can declare the array with the following.

```
typedef enum { WARNING_FLAG, OK_FLAG, SYSTEM_IDLE_FLAG,
               SYSTEM_BUSY_FLAG }
BooleanFlags;
Boolean booleanState[NUMBER_OF_BOOLEANS];
```

Now to manipulate the model, I set a Boolean to TRUE with the following.

```
booleanState[OK_LED] = TRUE;
```

This can be encapsulated in a function call.

```
setBooleanTrue(OK_LED);
```

By hiding it in a function, the programmer manipulating the model no longer needs to be aware of the structures used to implement the model.

It is important to note that this array is not memory-mapped to the physical LEDs. The preceding assignment has updated the model, but now the display is out of date. One way to keep the view up-to-date is to add code to the setBooleanTrue() function to update the display. This manipulation may be a simple write to a single bit, or a more complex mechanism may be in place if the LEDs are multiplexed.

Embedding this code to handle the view functionality in the setBooleanTrue() function works for simple cases, but it lacks flexibility. Almost the same code would be repeated in the setBooleanFalse() function and in the clearAllBooleans() function. It also makes manipulation of the model and the view very tightly coupled, because a change in the way the view is updated would cause changes to the functions called to update the model. A better arrangement is to write a single function that maps the whole Boolean array to the view. I will call this function refreshLeds(). This function now holds all the view functionality in one place. The implementation of this function is not shown, because it depends on the hardware configuration of the particular system. The function loops through the booleanState[] array and turns each LED on or off by toggling a bit or by writing some data to a chip that controls multiplexing of a set of LEDs.

It may seem inefficient to update all the LEDs each time one Boolean is changed, but as the examples get more complex, the advantages of maintainable code start to outweigh the performance price. Later, you will also see that many changes can be made to the model, followed by a single refresh. This is more efficient than updating the display at the end of each function that manipulates the model.

The application code or the controller can manipulate the model at will. The application is aware that it is updating Boolean values, but it need not be aware that the values are represented by LEDs on the user interface. If an alternative view were constructed, such as displaying the flags on a text display, then another refresh function might be written. In fact, both views can be visible simultaneously, which would involve no changes to the model or the application code. This is an example of encapsulation, because the view has been made independent of the rest of the application.

When the view needs to be brought up-to-date, refreshLeds() can be called. This function is not necessarily called after every change to the model. Most of the time, the device responds to events. Whether the event is generated by the user, via the controller, or by another part of the device, the duration of the processing of the event is often short in terms of human response times. The event might be fully processed in several milliseconds. Refreshing all LEDs at the end of each event, rather than from the end of each change to the model, is then acceptable to the user. Because all events pass through the controller, there is a point in the controller code where I can put the call and be sure that it will be handled for all events.

At this stage, the code to handle each event can simply call the functions to manipulate the model, knowing that the view will be updated when the event processing is complete. If some event takes an exceptional amount of time to complete, it may call SetBooleanTrue(SYSTEM_BUSY_FLAG), followed by an explicit call to refreshLeds() to update the display before the event has completed. But such out-of-turn calls to refreshLeds() would be rare.

> **Do not update the display at the first possible opportunity. Wait to update the display based on a number of changes to the model.**

2.3.1 Optimizations of the Refresh Function

If a change to the model can be recorded in some compact form, it allows further optimizations when refreshing. For a vertical column of LEDs used as a bar graph, it may be possible to record the change in height of the column between refreshes. Each successive refresh could record the current height so that the following refresh can compare the new model to it. Only the changed LEDs need to be updated, to make the column taller or shorter.

Additional complex refresh algorithms are seen in Chapter 5, which deals with refreshing graphics displays.

2.4 Layered Messages on a Small Text Display

You may want to show only a certain amount of the model to the user. Sometimes you do not want to overwhelm the user by showing a large amount of information at once. Or perhaps the display hardware does not allow all the information to be displayed at once. This issue often arises with small text windows. The HD44780 is a typical controller for small (four-line by 20-character) displays. Labrosse (1995) describes how to program this device at the register level.

One problem with custom displays is that the basic control of the output devices is often so simple that the programmer is lured into a false sense of security. The programmer who can put text on a 20-character display with a few lines of code assumes that displaying the right message at the right time can be only a matter of tagging some control logic around a trivial device driver. However, the demands at a higher level may be complex if the messages come from a variety of sources. Some text messages are in response to user actions. Alarm conditions may occur at any time. Possibly, there is a clock running which should be displayed if there is nothing more important to display.

This can be troublesome if the application has to check what is currently on the text window and then decide if the new message is more important. If a more important message is on the window, you do not display the new message immediately. However, once that message is removed — say, the alarm condition goes away — then you want your less important message to be visible. One way of solving this problem is to provide an interface to the text window that provides layers. For each line of the text window, there may be several layers. The topmost nonempty layer is the one that is visible when the display is refreshed. Any layer can be written to at any time without the application having to worry about whether that message should appear now, later, or not at all.

The layers for each line are independent. You may display the fifth layer of the first line and the third layer of the second line. There is no need to equalize the number of layers on each line, but I have done so in this example to simplify the implementation.

This text display manager shows another advantage gained when the display is not updated directly, but is deferred to a refresh function. It is often easier to decide what should be displayed when mapping the model to the view in the refresh function rather than at the time the model is updated.

Figure 2.4 shows a set of layers for a two-line display. There are three layers. I will call a single line in a layer a *slot*. Each of the six slots has a purpose. The application level cares only about which slot it is writing to. It does not care where that slot is located among the layers. For example, the time-of-day slot is updated every second, regardless of whether it is visible. If the time-of-day slot is changed to the second line so that it is visible at the same time as an urgent warning, the allocation of slots changes, but the application code should not change. In this example, slots are given names by way of a set of #defines. This makes the reallocation of slots simple.

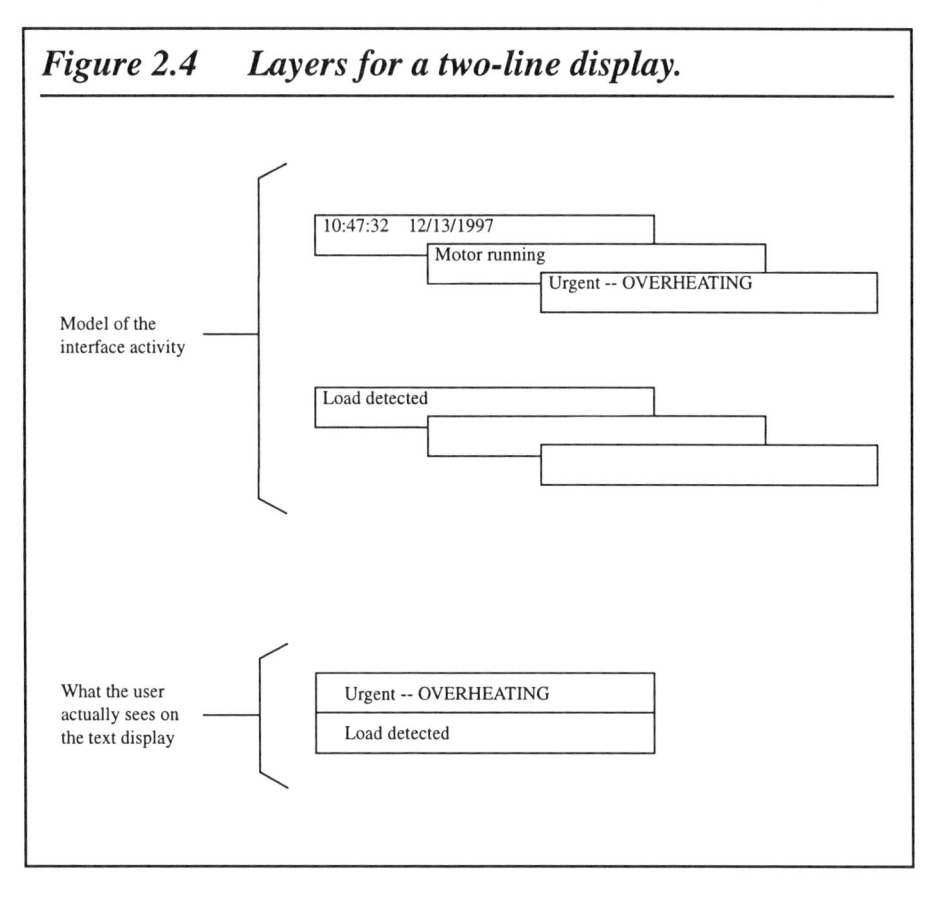

Figure 2.4   Layers for a two-line display.

On the top line are the number of messages that have been requested. The "Urgent OVERHEATING" message is to the front, so it will be displayed. The time-of-day slot is at the back, so it would only be visible if the "Urgent OVERHEATING" message and the "Motor running" status message were not present.

On the second line, the "Load detected" message is on the back layer. However, the first two layers are blank, so the "Load detected" message is visible.

In a more realistic case, many possible messages may be used for each slot. A slot may be associated with a category of messages, such as one dedicated to the motor status. In the example program on the companion disk, the "Motor running" message could be replaced by the "Motor stopped" message. Because it is replaced by overwriting the same slot, the "Motor running" message will not reappear if the "Motor stopped" message is removed.

This scheme adapts easily if a larger display becomes available. The slots may be spread over a larger number of lines, making more information available simultaneously. Another possibility is that only a small number of lines are available via the device's display, but all slots may be printed to the serial port, which allows a PC or an optional accessory with greater display ability to show all activity.

This scheme does not allow for splitting horizontal lines, as you may wish to do on a full-size text screen. For that sort of application, you probably need a more powerful set of tools that provides full text windowing.

2.4.1 The Text Display Manager Module

The text display manager (TDM) module enables you to manage a text window of arbitrary size. This module is independent of the screen's dimensions and physical implementation, which makes it possible to reuse it in your code with no changes to the tdm.c file itself.

The following header file shows the functions visible to the rest of the system.

```
#ifndef TDM_H
#define TDM_H
/*
tdm.h
*/
/*
This text display manager allows the user to create a manager for a text
window, so long as a display function can be provided that conforms to the
prototype:
void display(int lineNumber, const char *string);
where lineNumber starts at 0 and string the text to be printed
on the display, and is null terminated if shorter than lineLength.
*/
typedef void (*DisplayFunction)(int lineNumber, const char *string);
```

```
/*
By making the type visible here without a full definition I have implemented
an opaque type. The caller can manipulate pointers to the structure but has
no access to the contents.
*/
extern struct tdmStruct;
typedef struct tdmStruct TextDisplayManager;

TextDisplayManager * TDMcreate(int lineLength, int numberOfLines,
                               int numberOfLayers, DisplayFunction display);
void TDMrefresh(TextDisplayManager *TDMptr);
void TDMset(TextDisplayManager *TDMptr, int lineNumber, int layer,
            char *string);
void TDMclear(TextDisplayManager *TDMptr, int lineNumber, int layer);
char * TDMgetSlot(TextDisplayManager *TDMptr, int lineNumber, int layer);

#endif /* TDM_H */
```

The tdmStruct, which is typedefed to the name TextDisplayManager is shown here, as it appears in tdm.c.

```
struct tdmStruct {
    int lineLength;
    int numberOfLines;
    int numberOfLayers;
    char ***layers;
    DisplayFunction display;
};
```

This structure can be accessed only from within the tdm.c module. The tdm-main.c module can manipulate pointers to the TextDisplayManager, but it cannot access the internals. This helps improve encapsulation and is often referred to as an *opaque type*.

Avoid putting definitions in header files unless absolutely necessary. Keep them local to the .c file.

The layers field is a char ***. This is a pointer to a pointer to a string. The TDMcreate() function allocates all the space required to implement this two-dimensional array of strings. Once the space is allocated, it is initialized by setting each string at zero length. Later, any string placed on any layer will be copied into this space. If TextDisplayManagers were to be created and destroyed dynamically while the program is running, you would need a matching TDMfree() function. This function would free all the space allocated in TDMcreate(), to avoid memory leaks. I do not provide that here because I assume that the display will be in use until the program exits.

The `DisplayFunction` is a pointer to a function capable of displaying a line of text on some kind of display. The pointer to a function is stored in a field of the `TextDisplayManager` so that it can later be called from the `TDMrefresh()` function. It represents the interface to the physical layer.

```
TextDisplayManager * TDMcreate(int lineLength, int numberOfLines,
                               int numberOfLayers,
                               DisplayFunction display)
{
    int i; /* used to index the lines */
    int j; /* used to index the slots */

    TextDisplayManager *TDMptr;
    TDMptr = malloc(sizeof(TextDisplayManager));
    assert(TDMptr != NULL);
    TDMptr->lineLength = lineLength;
    TDMptr->numberOfLines = numberOfLines;
    TDMptr->numberOfLayers = numberOfLayers;
    TDMptr->display = display;
    /*
    The assert here checks that malloc did not fail and return null
    */
    TDMptr->layers = malloc(numberOfLines * sizeof(char **));
    assert (TDMptr->layers != NULL);
    for (i=0; i < numberOfLines; i++)
    {
        TDMptr->layers[i] = malloc(numberOfLayers * sizeof(char *));
        assert (TDMptr->layers[i] != NULL);
        for (j=0; j < numberOfLayers; j++)
        {
            TDMptr->layers[i][j] = malloc(lineLength*sizeof(char));
            assert (TDMptr->layers[i][j] != NULL);
        }
    }
    /* Initialise each of the slots to an empty string */
    for(i=0; i < numberOfLines; i++)
    {
        for(j=0; j < numberOfLayers; j++)
        {
            TDMptr->layers[i][j][0] = '\0';
        }
    }
    return TDMptr;
}
```

Because `TextDisplayManager` is an opaque type and its size is not known to compilation units outside `tdm.c`, you have lost the ability to declare them statically or on the stack. Using the heap is not acceptable because heap fragmentation eventually causes an allocation to fail. The technique I often use is to allocate structures such as the TDM on the heap, but I create such structures only at start-up. No access to the heap is made once the system is initialized. In this way, you can be sure that if the system is going to fail a heap allocation, it will do so at start-up and the problem will be caught and corrected during the design phase. If you do want to use memory dynamically, the pointer returned from `TDMcreate()` cannot simply be passed to the `free()` function. If it were passed directly to `free()`, the memory allocated for the lines and layers would be leaked. If you want to use these structures dynamically, a `TDMdestroy()` function would be necessary. This function would free all the allocated space pointed to by the `TextDisplayManager` structure, and then it would finally free the structure itself. Memory management is discussed further in Chapter 5.

The `TDMset()` and `TDMclear()` functions allow the creator of the `TextDisplayManager` to change the text in any of the layers. Note that the text is copied in, it is not just a pointer, so the caller may change or delete its copy of the text. These are the functions that manipulate the model.

```
void TDMset(TextDisplayManager *TDMptr, int lineNumber, int layer,
            char *string)
{
    char *startOfSlot;
    /*
    I could drop this assert if I were happy to lose characters
    off the right-hand side of the display.
    */
    assert(strlen(string) <= TDMptr->lineLength);
    startOfSlot = TDMptr->layers[lineNumber][layer];
    strncpy(startOfSlot, string, TDMptr->lineLength);
}
void TDMclear(TextDisplayManager *TDMptr, int lineNumber, int layer)
{
    char *startOfSlot;
    startOfSlot = TDMptr->layers[lineNumber][layer];
    *startOfSlot = '\0';
}
```

The zero-length string set in `TDMclear()` is a special case that makes the line transparent rather than a line of spaces that would hide any text on layers behind.

When I make changes to the model, the `TDMrefresh()` function implements the view. It maps the model to one string per line of text. These strings are displayed using the display function stored in the `TextDisplayManager` structure. This function pointer gives us access to the physical layer. In the demo program, it points at a function that merely prints the strings on a certain part of the PC screen. In a more realistic embedded system, it may drive a controller chip for an LCD screen. If the program were capable of driving a number of text displays in different versions of the product, changing this function pointer would be an efficient way to switch between physical layer implementations at run time. If you do not require that kind of flexibility, the program may be simplified by removing the pointer to a function and replacing it with a hard-coded call to one display function.

```
void TDMrefresh(TextDisplayManager *TDMptr)
{
    int i; /* used to index the lines */
    int j; /* used to index the layers */
    Boolean lineDone;
    char *startOfSlot;

    /* Loop through the lines */
    for(i=0; i < TDMptr->numberOfLines; i++)
    {
        lineDone = FALSE;
        /* loop through the layers, searching for the first layer
           that has a non-empty string */
        for(j=0; (j < TDMptr->numberOfLayers) && !lineDone; j++)
        {
            startOfSlot = TDMptr->layers[i][j];
            if (startOfSlot[0] != '\0')
            {
                lineDone = TRUE;
                TDMptr->display(i, startOfSlot);
            }
        }
        if (!lineDone)
        {
            /*
                There is nothing to be displayed on this line so display
                blanks to cover anything that may have been on the line
                from a previous refresh. I know that startOfLayer is an
                empty string at the moment, and the display function
                will fill out the unused spaces with blanks.
            */
            TDMptr->display(i, startOfSlot);
        }
    }
}
```

The `tdm.c` module does not put names on the layers. In the actual use of this module, it is preferable to name slots rather than using numerical indexes, because the desired positions for certain categories of text may change during the life of a project. In the example program, the slots are named in `tdmmain.c` as follows.

```
#define URGENT_LINE 0
#define URGENT_LAYER 0

#define MOTOR_LINE 0
#define MOTOR_LAYER 1

/* TOD stands for Time Of Day */
#define TOD_LINE 0
#define TOD_LAYER 2

#define LOAD_LINE 1
#define LOAD_LAYER 2
```

Now a `TextDisplayManager` can be created and the motor status slot can be set with the following.

```
TextDisplayManager *tdmPtr;
tdmPtr = TDMcreate(20, 2, 3, displayText);
TDMset(tdmPtr, MOTOR_LINE, MOTOR_LAYER, "Motor Running");
```

2.5 Conclusion

You have seen how the MVC paradigm can influence the design of some small examples. It could provide a model against which to analyze existing systems in a reverse engineering role. The subsystems defined in this chapter are found in many existing systems that were not designed to meet an MVC decomposition. In some cases, the subsystems are provided by third-party vendors.

Although the MVC examples provided here show only a single instance of each part, far more complex arrangements are possible and sometimes necessary. On a windowed system, there would typically be one model, one view, and one controller per window. This allows the windows to operate independently of each other. The only points of interaction are where the events are guided to the appropriate controller and where the refresh routines must cooperate to ensure that one window does not corrupt another.

Several other programming paradigms may be applied to interactive systems (Gram et al., 1996). Such systems, which go beyond the scope of this book, are generally variations on the MVC paradigm and break the software into a greater number of layers. For the purposes of this book and for most typical systems, the MVC provides a foundation for well-structured user interface software.

Chapter 3

Scheduling and Managing User Events

Chapter 2 discussed the model-view-controller (MVC) paradigm used to decompose a system. Serious discussion of the controller was not handled in that chapter, because it is better discussed in tandem with queuing and tasking, which I will cover in this chapter. When an event has been successfully detected, it may not be possible to respond to it immediately, so the response may have to be scheduled for execution at a later time. Queues and tasks allow us to schedule and control such asynchronous activity.

This chapter makes use of the functionality provided by a real-time operating system (RTOS). The RTOS used in the examples is µC/OS (Labrosse, 1992), but its functionality is similar to that provided by any other RTOS. If you are not familiar with the use of an RTOS, you may want to read Appendix A before proceeding.

3.1 User Events on the Desktop

With a desktop operating system, much of the interpretation of events has been handled for you. Some simple function allows you to request the next key pressed by the user. Any queuing, debouncing, or mapping from the codes returned by the keyboard to standard ASCII characters has already been performed. If the system has a graphical user interface (GUI), mouse events may have already been associated with the object the user selected. In some cases, the control is passed to the graphics environment via a function called something like `ProcessUserEvents()`. That function reads the event queue and calls the appropriate function for each event as it is popped off the queue. Much of the flow of control is hidden from the programmer, as it should be. There is no reason for the programmer to know how the key or mouse events are read, filtered, or queued by the system.

However, the embedded programmer is often faced with hardware designed in-house that cannot be integrated with a commercial operating system. Many of the principles implemented by the operating system or by GUI toolkit vendors are the same as those you can implement in your own embedded system. The principles are the same whether the interface is as complex as a full GUI or as simple as a digital watch.

3.2 Polling versus Interrupts

At the lowest level, events are detected by the hardware. Software must detect any change in order to handle the event. There are three basic methods.

- Polling from the main flow of control
- Timed polling
- Interrupts

3.2.1 Polling from the Main Line

Polling from the main flow of control means reading the hardware once every time around the main loop.

```
void main(void)
{
    initialize();
    while (1)
    {
        e = readEvent();
        if (e is a valid event)
        {
            processEvent(e);
        }
    }
}
```

This technique works in simple cases, but if processing an event takes longer than the time for an event to occur, some events may be missed. Events cannot be queued to be processed later. This gets more difficult if there are several kinds of events that you need to prioritize. Because of these restrictions and because of its simplicity, this approach is not discussed any further.

3.2.2 Timed Polling

Timed polling involves reading the hardware on a regular, timed basis. If an event is found, it is queued. The main flow of control then reads the queue after each event has been processed. Polling of the hardware could be triggered by a timed interrupt. If an RTOS is available, it is preferable to have an RTOS task wake up from a sleep every time a read is due. So now there are two loops: one for the polling task and one for the task processing the events. Each loop runs in a separate task, so they are executed in parallel.

```
void keyPollTask(void)
{
    while (1)
    {
        e = readEvent();
        if (e is a valid event)
        {
            enqueue(e);
        }
        sleep(poll_period);
    }
}
void eventProcessingTask(void)
{
    while (1)
    {
        e = dequeue();
        processEvent(e);
    }
}
```

This arrangement allows the user to type ahead. The extra events sit on the queue until the event-handling loop is ready for them. The flow of control can be seen in Figure 3.1. The event e2 was detected before e1 had been completely processed. If the poll had not taken place before the processing of e1 was complete, the event could have been missed completely. The polling period must be chosen so that no events are missed. At the same time, the higher the frequency of polling, the more CPU cycles are spent checking for events, even when no events have taken place. Missing events is less likely if the hardware latches an event so that it can be detected an arbitrary period of time after the occurrence. The latched data is then cleared by software so that the hardware is ready for the next event. This is only a partial solution because during the period that data is in the latch waiting to be read, no further events can be detected. The latch has effectively provided a queue of length one.

If there are delays in the hardware during the `readEvent()` function, the control can switch to the event being processed during those delays. A delay may be inserted to allow an output signal to settle before reading the input. These context switches are not shown in Figure 3.1.

If you are concerned that the polling interval you have chosen may miss events occasionally, simply run the system with the polling set at half the design frequency. If no events are missed, you can trust the design frequency. Such a check is worth the effort, because operators will get quicker at using the device over the years, possibly quicker than the testers. As well as decreasing the frequency, it is also useful to increase the polling frequency during testing. Some motion detection devices may generate an event on every poll. This stream of events may be in danger of flooding another part of the system. Increasing the polling frequency stress-tests such conditions.

3.2.3 Interrupts

To use interrupts, the event-detecting circuit must be wired to one of the processor interrupt lines, while the polling methods require only that the event-detecting circuit be readable, usually using the address and data lines. An interrupt, unless disabled, immediately switches the flow of control to the interrupt service routine (ISR). Similar to timed polling, the ISR queues the event until the main flow of control has time to process it. Figure 3.2 shows the timing diagram for the same events with interrupts.

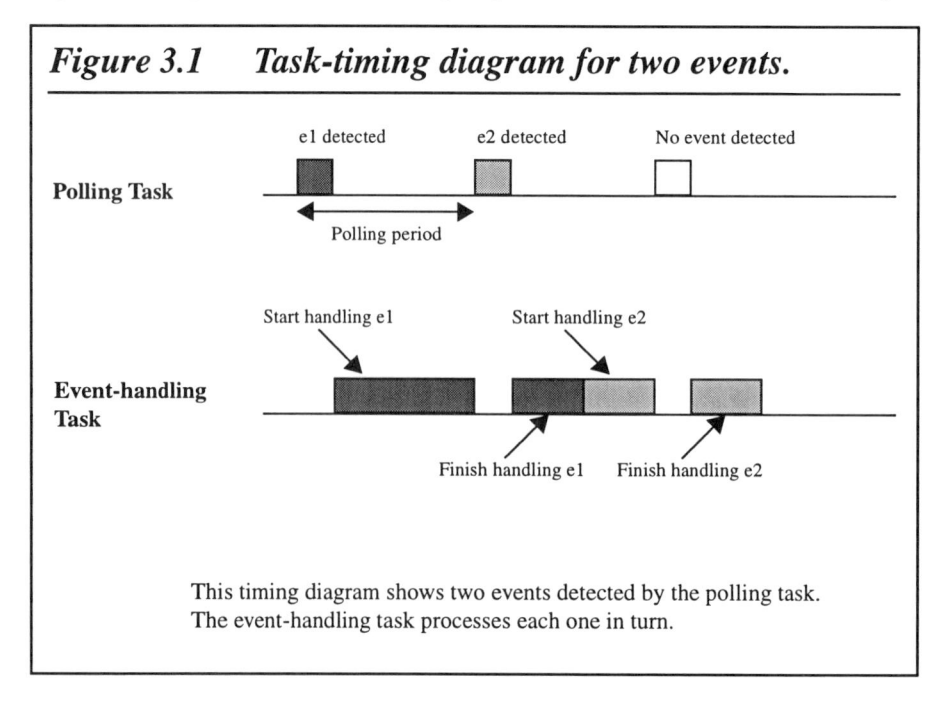

Figure 3.1 Task-timing diagram for two events.

This timing diagram shows two events detected by the polling task. The event-handling task processes each one in turn.

Note that there is no interrupt when there is no event. In this case, the time between the detection of e1 and e2 is not predefined. Interrupts are useful when the event is very brief and may be missed by a polling routine.

The interrupt mechanism can be used without an RTOS. Effectively, the interrupt supplies one thread of control while the main flow of execution provides another. Obviously, the interrupt mechanism can still be used if there is an RTOS available, but note that it is not possible to place the interrupt's thread of control into the priority order dictated by the RTOS. This may be important if you do not want the keyboard-scanning routines to interfere with other high-priority work that the processor must perform. Disabling the interrupt during some high-priority sections of the code may be an option, but this can increase the complexity of the code.

The overhead associated with polling, when there is no input present, might make you think that the interrupt-driven system is more efficient. This is not always the case. The average overhead in the interrupt-driven system is lower, but the average can be a misleading metric. Consider a keypad that is polled every 10 milliseconds (ms). It takes 0.1ms to poll the keypad when there are no keys selected and 0.3ms to perform the poll when there is a key present. So the overhead associated with the keypad is between 1 percent and 3 percent of the total CPU bandwidth. It is likely to be 3 percent when the system is doing most of its work. This is because once the user presses a key, he is likely to keep it depressed for several polls. Say, the user holds down a button for 300ms. During that time, the event is detected and enqueued to the event-handling task.

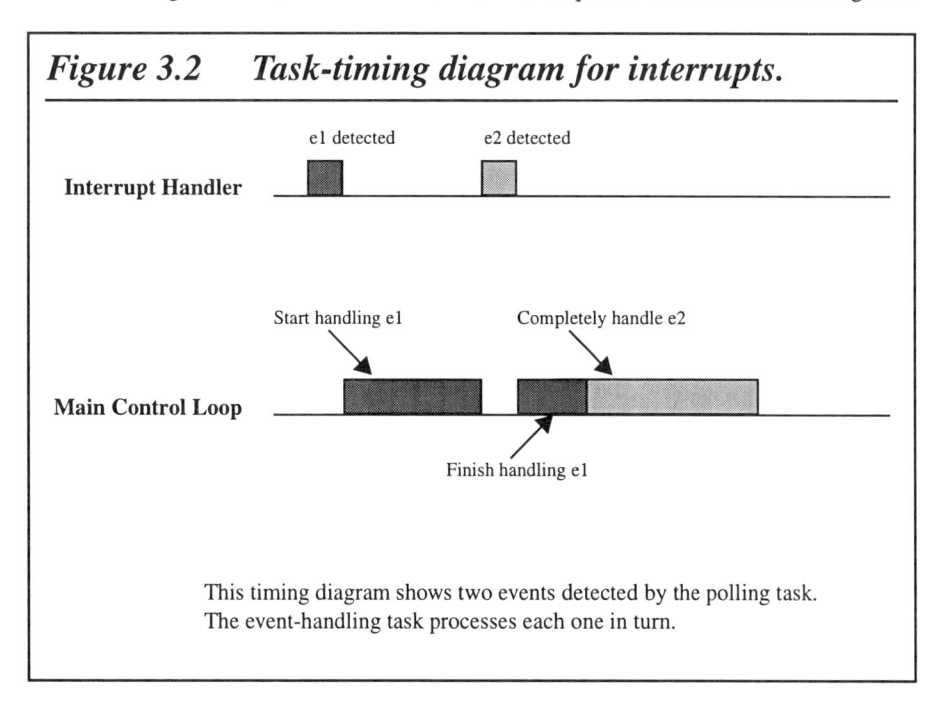

Figure 3.2   Task-timing diagram for interrupts.

This timing diagram shows two events detected by the polling task. The event-handling task processes each one in turn.

The event-handling task processed that event. Say that event takes 250ms. The event is complete, but during its processing, the overhead from the keypad is 3 percent because the user kept his finger on the key. This is not unrealistic, because most people move their fingers slowly relative to computer speeds. A similar scenario arises with motion devices such as a mouse or a dial. While one movement is being processed, it is quite likely the user has made another movement. The events from a motion device arrive in bursts. For these bursts of activity, the timed polling approach has another advantage. The change between two consecutive reads reflects the speed at which the device is moved. This makes it easier to apply rate-sensitive algorithms to the response. This is not true of an interrupt-driven approach because the time between successive reads varies, and measuring that time may be nontrivial.

Now, look at the response in an interrupt-driven system. If the key input is not filtered electronically, the interrupt may have to be debounced by disabling or ignoring the interrupt for a certain period. Sometimes, it is not possible to disable the interrupt without the risk of missing other events. If the interrupt is not disabled, many interrupts may take up a large percentage of processor time at the exact moment an event needs to be processed. If you can disable interrupts, you need to find an appropriate point in the code at which the interrupts can be reenabled. If you wait until the event is processed, you risk missing a second event.

If the period for which you disable the interrupts is the same as the poll period, the overhead when the system is busy is the same for both designs. Picking the same figure for the poll period and for the interrupt disable time is reasonable, because this is the amount of time that can pass without missing an event. The performance during the rest of the time is often irrelevant. If there is no system activity on the user interface, the processor responsible for the user interface will have spare CPU cycles available.

For highly responsive systems, interrupts are the only way to detect events, but on a user interface, you should always give timed polling serious consideration. The overhead is bounded and predictable, the hardware is less complex, and most important, the software is easier to write and, therefore, to debug.

3.3 Queuing

Whether the input event is detected in a polling task or in an interrupt, the request has to be queued. If you have an RTOS, it provides a queue structure. The enqueue and dequeue operations are atomic, avoiding any problems caused by simultaneous access. If you do not have an RTOS, you can implement the queue by using an array as a circular buffer. You also need some means of locking that data to ensure that it is not read while it is being written. Disabling interrupts while the queue is being read is the simplest way. Within the interrupt routine itself, other interrupts have to be disabled while the queue is being written to protect against nested interrupts causing overlapping writes to the queue.

Scheduling and Managing User Events — 33

The length of the queue depends on the nature of the input. In desktop systems, a lot of type-ahead is appropriate, allowing the user to type commands or sentences. Most embedded systems do not need more than one-key type-ahead. If key-up as well as key-down events are detected, you may want to allow a longer queue. Motion devices that generate a stream of events also require a longer queue. Once the queue is full, most systems function if the following events are thrown away. Users soon realize that they have limited type-ahead (or move-ahead, in the case of a motion device). If you discard events, you must ensure that events do not depend on each other. If a key-down occurs, the logic of the controller may be upset if the corresponding key-up never occurs. If this is the case, the queue must be sized to cope with the highest possible input rates and never lose events.

A general-purpose queue contains elements of a fixed size. µC/OS allows only a single void pointer as the element of the queue. In the example that follows, I maintain a separate array to hold the data and point to it from the µC/OS queue.

The structure placed on the queue defines a protocol between the physical layer and the controller. Consider a control system with the display shown in Figure 3.3, which illustrates the screen layout when running the keyq example program on the companion disk. The user can view and set three parameters: temperature, pressure, and flow. The LED lights when the key next to it is selected. The corresponding setting then appears in the numeric display, and the user can adjust the values using the arrow keys. If the display is not touched for 10 seconds, it times out, the numeric display goes blank, and all LEDs are turned off. The display is also capable of displaying two different alarms: high temperature and high pressure. So the task managing the

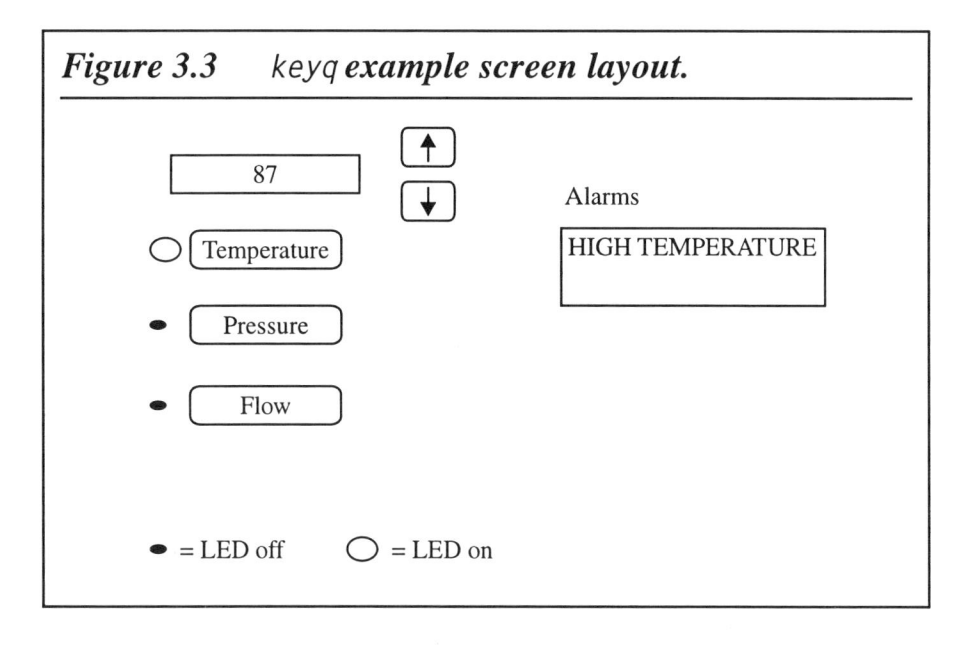

Figure 3.3   keyq example screen layout.

display can handle three types of event: key events, alarm events, and timeouts. I can place these three types of events on the queue using the following structures.

```
/*
Key Mapping
*/
typedef enum  { TEMP_KEY, PRES_KEY, FLOW_KEY, UP_KEY, DOWN_KEY,
                NO_KEY} KeyValue;

typedef enum { HIGH_TEMPERATURE, HIGH_PRESSURE } AlarmType;
typedef struct {
    AlarmType type;
    Boolean value;
} AlarmStruct;

/* Packet Structure */
typedef enum {KEY_PACKET, ALARM_PACKET, TIMEOUT_PACKET} PacketType;
typedef struct {
    PacketType type;
    union {
        KeyValue keyValue;
        AlarmStruct alarm;
    } data;
} QPacket;
```

By storing the data for the different types of event in a union, I can minimize the size of the packet I place on the queue. Notice that there is no data associated with the timeout. The fact that it occurred is sufficient information because there is only one type of timeout.

3.4 Is an RTOS Needed?

For many small projects, no RTOS is required. Many complex programs would be almost impossible to write without one. A lot of projects fall somewhere in between, leaving the developers wondering if the learning curve, and possibly a license fee, are worth the investment.

An RTOS provides the ability to manage time, so the more limiting and complex the time constraints of the program, the more likely you can benefit from an RTOS. User interfaces tend to be slower than many other parts of the software, and they do not need to react as fast as process control feedback loops or communications software. However, the processor that controls the front panel generally has other responsibilities. An RTOS allows the designer to limit the processor time that responding to user events may consume. On many devices, it is not acceptable to allow the user to halt some operations by repeatedly pressing on a key. Even if the key press is invalid and meaningless, handling it may consume processor time that should be spent on some other activity. It is possible to manage such timing issues by carefully structuring the code to provide for each of several functions that get processed in turn. However, this can lead to spaghetti code. If you have a function that takes a relatively large amount of time, you may have to place a function call in the middle of it, which passes control to an unrelated area of the software to allow it to meet some deadline. Such out-of-place calls are a sure sign that it is time to invest in an RTOS.

Because an RTOS supports multiple stacks and multiple threads of control, it allows one event to be halted while a more important event is processed. The task that has been halted can resume from the point at which it was before the context switch. Interrupts provide a similar ability, but with some limitations. An interrupt requires an external event. Sometimes, a task is driven by an internal event, such as a queue-not-empty event. Generating an interrupt to reflect the internal event may be difficult and may require extra hardware. Interrupt handlers work well when the work they must perform can be restricted to a small time-slice. If an event triggers a function that will take several seconds to process, interrupts will prove unsuitable.

One approach to the task decomposition is to dedicate one task to the user interface, while a number of other tasks manage the other responsibilities of the processor. This works if user events can always be processed in a very short time. A fast processor may mean that the computations resulting from the event can be performed quickly, but interfacing with other hardware may slow things down considerably. This leads to timing requirements at two levels, which in turn leads to at least two tasks. Assume that you are required to detect all key events and that the mechanical characteristics of the key dictate that the shortest period during which the key event is detectable is 50ms. If key events are polled, then a task must run every 50ms, or faster, depending on whether the events are latched by the hardware. The polling itself is generally a quick job and does not cause any other deadlines to be missed. The second requirement is that while the event is being processed, a far longer period is needed. Events may vary dramatically in the amount of time they take to be processed, but this variation must not interfere with the performance of other tasks with real-time deadlines. There is a danger that a continuous series of user events may halt any task with a lower priority than the event-processing task.

Therefore, you generally want the polling task to have a higher priority than other tasks that might delay the polling long enough that a user event may be missed. The event processing task should run at a lower priority than tasks that cannot afford to be delayed by user events. You will see another reason for the relative priorities of these two tasks after I examine the tasks in the keyq example in the next section. The keyq example is so small that it would not normally require an RTOS, but it is used here to illustrate some important issues.

Although this analysis attempts to give you enough information to decide if an RTOS is required for a particular project, the hardest part by far is to predict the timing behavior of your program before it is written. This is one area in which there is no substitute for experience.

3.5 The Tasks

Using the queue defined previously as the communications protocol, I will divide the `keyq` system into three tasks. The first is the `keyPoll` task, which polls the keypad and sends a timeout if no keypad activity happens for 10 seconds. This task implements the physical layer. The second is the monitor task, which raises alarms if the process goes outside predefined temperature and pressure limits. The third task is the `mvc` task. This task contains the model's data and performs all the work of interpreting events (the controller) and refreshing the display (the view). Figure 3.4 shows the various subsystems.

The interpretation of an event in the `keyPoll` task is independent of context. You are merely interpreting which physical key was pressed. Once the event has been passed to the controller, the controller can interpret that event using context supplied by the model and the view. The `pollKey()` function is called to read the current character, if there is one. In the example code on the companion disk, this function merely

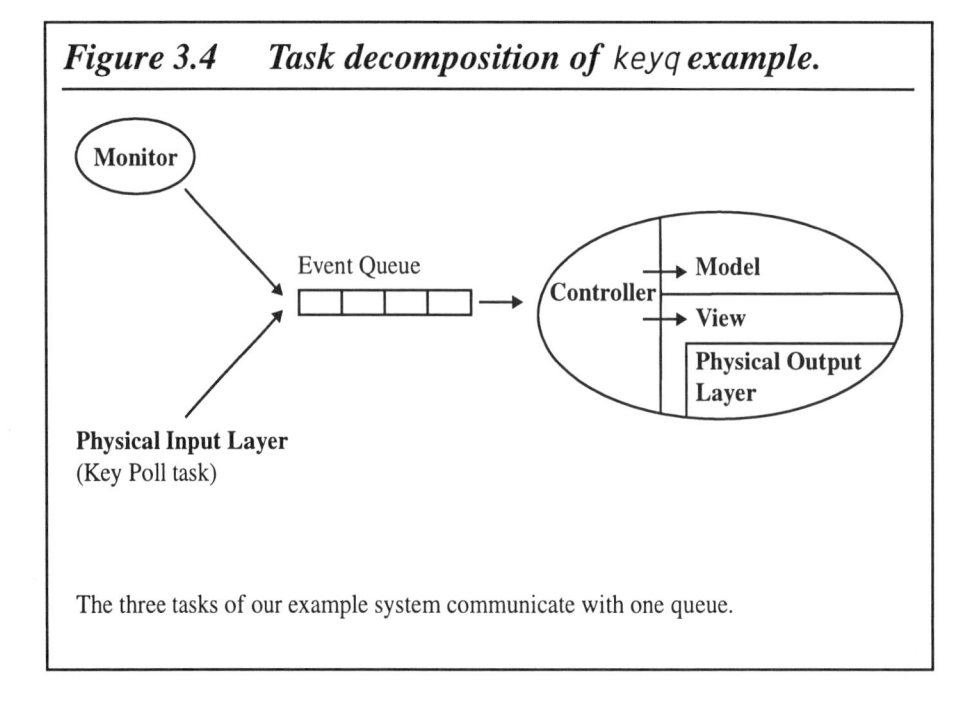

Figure 3.4   Task decomposition of `keyq` example.

The three tasks of our example system communicate with one queue.

uses the DOS functions kbhit() and getch(). In a real embedded system, the hardware would be accessed rather than using kbhit() and getch(). Note that the delay at the bottom of the loop is not equal to the poll period. The poll period is actually the delay at the bottom of the loop plus the time it takes for one execution of the loop.

```
void far keyPollStart(void *data)
{
    KeyValue key;
    int timeoutCounter = 0;
    data = data; /* avoid compiler warning */
    while (1)
    {
        key = pollKey();
        if (key != NO_KEY)
        {
            enqueueKey(key);
            timeoutCounter = 0;
        }
        /*
        Decide if no key has been pressed for
        approximately 10 seconds
        */
        timeoutCounter++;
        if (timeoutCounter == 6*10)
        {
            enqueueTimeout();
        }
        /*
        A delay is necessary here to ensure that this task does not
        continuously poll, hogging the CPU. Note that 0.16 seconds
        is fast enough to read the keys because they have already
        been queued. Many keypads can have a complete key event in
        0.05 seconds and if the keypad has not been polled in that
        time then it is too late - the event has been missed.
        */
        OSTimeDly(TICKS_PER_SECOND / 6);   /* delays 0.16 second */
    } /* end while (1) */
}
```

The calls to enqueueKey() and enqueueTimeout() place the appropriate event on the queue. The actual implementation is μC/OS-dependent because μC/OS supports only queue elements of type void pointer. For this reason, larger buffers have been associated with the queue in the enqueuing functions.

Once the data is on the queue, the controller must have a way to extract it. The infinite loop of the mvc task performs the first step of this interpretation.

```
void far mvcStart(void *data)
{
    UBYTE errCode;
    void *msg;
    QPacket *packetPtr;
    data = data; /* avoid warning */
    initDisplay();
    /* Initialise the model's data */
    GS_settings.temperature = 10;
    GS_settings.pressure = 20;
    GS_settings.flow = 30;
    while (1)
    {
        /*
        Wait for next event.
        Second argument is 0 meaning block until queue contains data.
        */
        msg = OSQPend(G_mvcQueue, 0, &errCode);
        assert (errCode == OS_NO_ERR);
        packetPtr = (QPacket *) msg;
        switch (packetPtr->type)
        {
        case KEY_PACKET:
            handleKey(packetPtr->data.keyValue);
            break;
        case ALARM_PACKET:
            handleAlarm(&packetPtr->data.alarm);
            break;
        case TIMEOUT_PACKET:
            handleTimeout();
            break;
        default:
            assert(FALSE);
        }
        refreshDisplay();
    } /* end while */
}
```

Note that this loop has no delay. The blocking point is the call to OSQPend( ), which does not return until there is some data on the queue. Thus this task consumes no CPU cycles while there is no activity on the queue. Once an event does appear on the queue, this task is ready to receive it.

The monitor task spends most of its time monitoring the hypothetical process driven by the device. Whenever the temperature or pressure exceeds the acceptable limit, the alarm is posted to the queue. In the example on the companion disk, I simply raise the alarms on a timed basis.

These tasks are actually configured and launched from the main function, but the details are highly μC/OS-dependent, so I will not dwell on them here. There is more information on task initialization in Appendix A.

3.5.1 Priority in a Tasking System

It is important for the `mvc` task that reads the queue to run at a lower priority than the tasks supplying the queue. That is because if the `mvc` task ran at a higher priority, it would prevent the `keyPoll` task from checking for a new event while the `mvc` task was processing an event, so a poll would not take place at the end of a poll period. The poll would be deferred until the `mvc` task completed processing that event. The chances of missing an event would be exactly the same as if polling from the main flow of control. It also means that the queue length would never be greater than one. Every time an event was placed on the queue, the `mvc` task would process that event and prevent any further activity on the `keyPoll` task. Compare Figure 3.5 with Figure 3.1 to see the difference in the sequence of actions. If you decide that Figure 3.5 is actually the preferred behavior for your application, the two tasks should probably be merged into one.

3.5.2 Allowing Multiple Tasks Access to the Display

In the example, the `mvc` task is the only task that has access to the model or to the physical display. In some cases, you may want different tasks to have access to the display. If some user events take time to process, you may want to perform other updates in parallel. Say you have a bar graph that is updated several times a second to

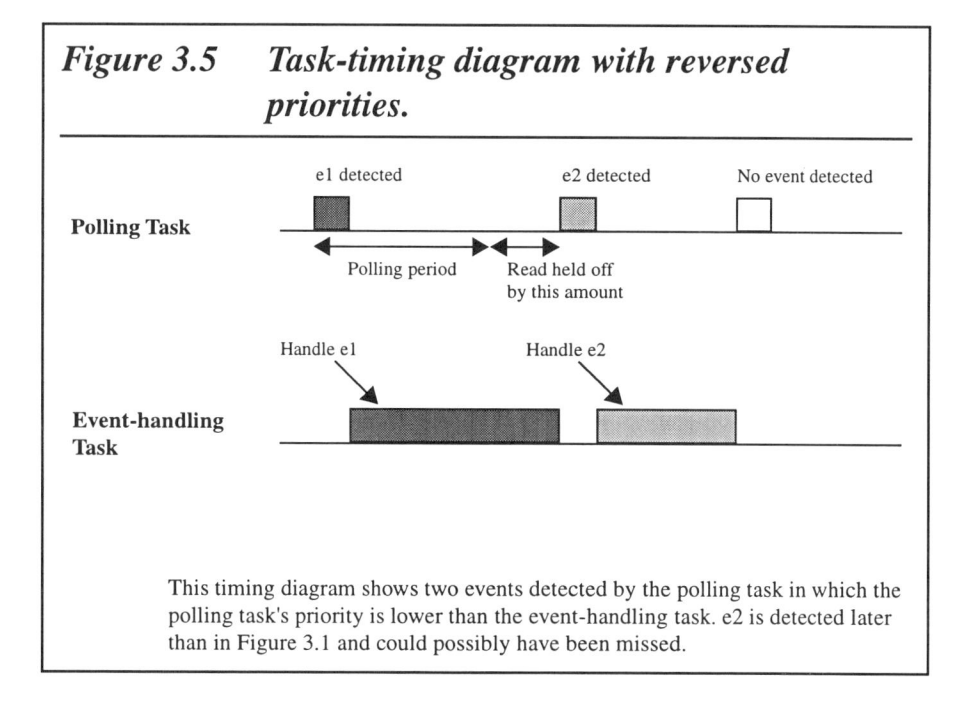

Figure 3.5   Task-timing diagram with reversed priorities.

This timing diagram shows two events detected by the polling task in which the polling task's priority is lower than the event-handling task. e2 is detected later than in Figure 3.1 and could possibly have been missed.

give real-time feedback of one of the monitored parameters. If some of the key-press events take more than a second to process, the bar graph appears to freeze briefly. A case could be made that one task should manage the key events while another updates the bar graph.

Breaking up the responsibility for updating the user interface raises a couple of issues. The hardware controlled by each task may be linked. The LEDs that make up the bar graph may be controlled by the same controller chip that controls other LEDs on the display. If two tasks simultaneously access that controller chip, the LED's display may become corrupted. Another example would be a graphics screen driven by a VGA controller. VGA controllers are not reentrant, so two tasks interleaving their access might lead to unpredictable results.

The hardware can be protected by a lock at the time of the update. Such locks, sometimes called resources, are supplied by most RTOSs. You can allocate one lock for each piece of the user interface hardware, or a single lock can be used for the whole front panel. Because all the updates to the user interface happen in a few refresh functions, it is straightforward to pick the places at which the hardware needs to be locked and unlocked.

A more complex issue is how to protect the model. Although the refresh functions localize hardware access, the model tends to be updated from several places, which can lead to a lot of locking and unlocking of resources. There is a small performance penalty here, but a much bigger cost is the increased complexity of the code. The only way to avoid the locks is to make each task own a portion of the model, and to make sure no task needs to access the model belonging to the others. This is not always practical.

Once several locks are used, the biggest danger is that the locks dictate that the tasks spend most of their time waiting for each other. When this happens, any improved response time for the short events is lost because the short event must wait for the longer event to complete and release its locks. In this case, the whole purpose of sharing the user interface over a number of tasks is defeated.

Before allowing a number of tasks to access the model and the user interface hardware, analyze the interactions to see if parallelism is really needed. In the bar graph example, will the user notice a slight pause in the graph's movement if he is concentrating on the response to the key he just pressed? If some of your events take more than a second to execute, examine the work they are doing and decide if some portion of it is unrelated to the user interface and could be spawned off to another task. The next section studies an example of this.

3.5.3 Breaking Up Long Events into Shorter Events

Suppose the user presses a button that starts a motor. I want to tell the user that the motor is running by lighting a LED. I want to provide the feedback only when the motor is running at a constant speed, and it takes a couple of seconds to reach that speed. The code that handles the key press could issue the commands to the motor and monitor the motor speed until it is stable. The user interface is now paralyzed for those few seconds while the motor accelerates because the mvc task cannot read the queue while it is monitoring the motor speed.

A reasonable solution is to pass the responsibility for controlling the motor to another task. The mvc task now makes the request to start the motor and then reverts to processing the next user event. The response to the key event may also include a key click or some other feedback so that the user knows that the key press was detected. The mvc task still has the responsibility for lighting a LED once the motor has reached its stable speed. This can be achieved via a second event that is queued from the motor-controlling task to the mvc task. Figure 3.6 shows the path of the event. In between item 2 and item 3, it is possible that the mvc task processed other events from the queue. The choice to use one queue to carry the key events and the motor-controller events was arbitrary; two queues could have been used, with the mvc task reading both.

Breaking up the processing of starting the motor this way means that the mvc task has two short jobs to process instead of one long one. In general, the user interface software is more responsive if it has many small events to handle rather than fewer large events.

3.5.4 A Task with a View

Another approach to the problem of allowing several tasks access to the display is to dedicate a single task to the view and the physical output layer. If there is a single

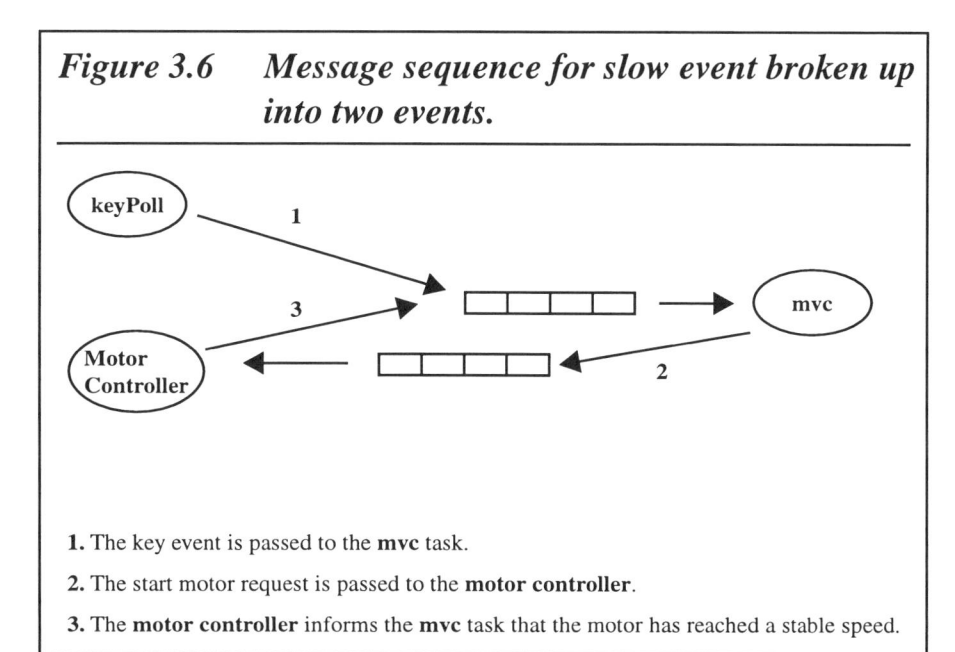

Figure 3.6 Message sequence for slow event broken up into two events.

1. The key event is passed to the **mvc** task.
2. The start motor request is passed to the **motor controller**.
3. The **motor controller** informs the **mvc** task that the motor has reached a stable speed.

queue to this task, any other task in the system can make a request to it. The overhead of placing the requests on a queue may prove comparable to the overhead of locking resources, and the design that evolves is a good deal easier to maintain.

Figure 3.7 shows this architecture. The monitor could now hold the model data for the alarms and instruct the view to display them directly. The request on the queue to the view would indicate which alarm to illuminate.

The model for the settings does not move, however, and still lives in the model/controller task. This could be seen as dividing the model into two portions, or it could be seen as two separate models. Even though some of the model resides in the monitor task, there is no need for any controller logic because none of the user events affect the alarm display.

The protocol to the view allows any elements on the display to be activated. The view task sees the LEDs simply as three LEDs and may not know about their relationship to the temperature, pressure, and flow settings. When the controller decides to display the temperature, the request to the view is to display the number 78. The view is not aware of the meaning of the number, and it cannot distinguish one type of setting from another. All the logic that allows settings to be manipulated is still in the model/controller task.

The view task is a convenient place to manage any timed activity that is purely visual, with no interaction with the model data, such as flashing LEDs. The view task would wait on its queue of requests, as well as wait on a timer that indicates when the state of the flashing elements should be toggled.

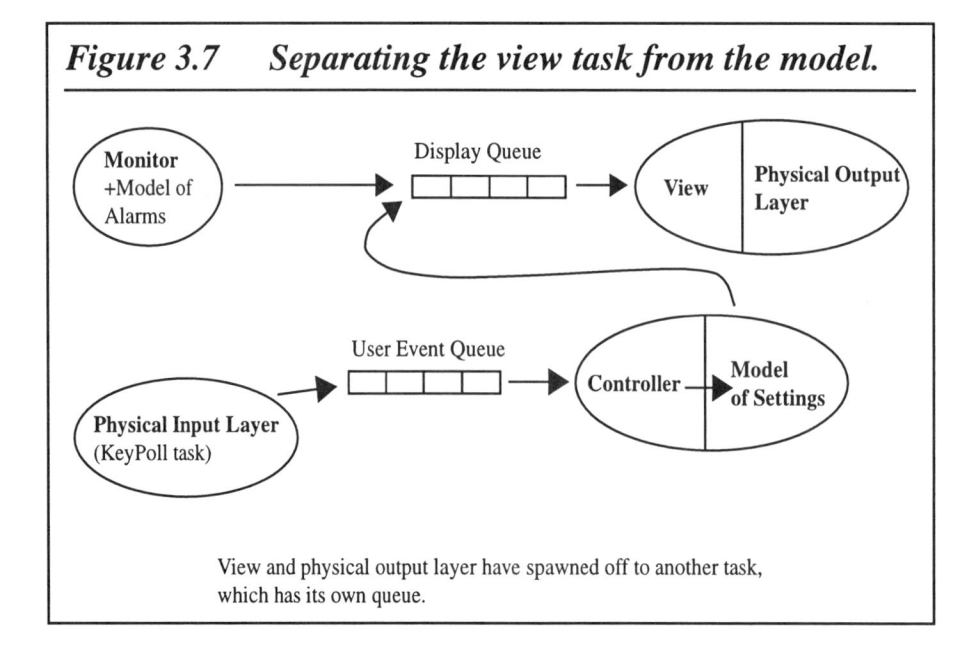

Figure 3.7   Separating the view task from the model.

View and physical output layer have spawned off to another task, which has its own queue.

Another possible use for the view task is multiplexing LEDs to reduce the number of control lines necessary (Labrosse, 1995). LEDs are often multiplexed in hardware by a dedicated LED controller chip, but multiplexing can also be performed in software to reduce the system hardware cost. The view task is a good place to manage this operation.

A queued protocol is trivial to convert to a protocol that can pass over a network or serial link. Now that I have separated the view, I have opened up the possibility of maintaining the model on a separate processor. This can be an attractive option for graphics systems in which the rendering of lines and arcs can be CPU intensive. One processor can be dedicated to managing the display, while all the application-specific control is performed on a separate processor. In such an architecture, the requests on the queue are at the level of requesting a line to be drawn between certain points, or an arc to be drawn with a given radius.

3.6 *Different Queues for Different Events*

As an alternative to the `QPacket` structure shown previously, each of the three types of data could be put on its own queue. This allows each queue to contain simpler structures. If you choose this route, there are several dangers. The RTOS you use may not give you a convenient way to block on multiple queues. If this is the case, it will be necessary to poll each queue in turn, adding to the delay between an event occurring and the event being processed.

Also, you must watch out for any dependencies between the order of events on one queue and another. Whether a key was pressed just before or after an alarm was annunciated may matter if the key's function is to disable or silence the alarm. A desktop example would be a mouse and a keyboard. If the mouse and keyboard events are handled on different queues and if both queues contain some elements when the task comes to read them, then there is no way of knowing which occurred first, the mouse event or the typing. If the typing is directed at the window containing the mouse, the typing may occasionally appear in the wrong window if the user works quickly.

> **If the order of events of differing types is related, keep them in the same queue.**

Even if you have several queues, each one is not restricted to one writer. Many tasks can write to a queue, and one task can write to many queues. On the reading side, the relationship is not so flexible. Even if your RTOS allows it, more than one reader of a queue is bad practice — events that pass through the queue may go to one of a number of tasks. The writer would not be sure which task received the packets.

Depending on the priorities of the tasks, one task may consume all the events and the others may get none. By giving each its own queue, this sharing of events can be more easily controlled. Figure 3.8 illustrates these configurations.

3.7 Queue Read-Ahead

Events from a motion detection device, such as a mouse, tend to occur in bursts. For example, if the mouse moves quickly from one side of the screen to the other, four or five events may appear in the queue before being processed. More typical motion devices in embedded systems are trackballs, dials, mechanical sliders, and thumb-wheels. These devices produce responses that can lead to the system response lagging behind the user's input. After the user finishes moving the control device, output continues to change until all events on the queue are processed. In order to reduce this lag, the reading task may look ahead on the queue to combine several events before processing them.

When a mouse event is popped off the queue, the next event on the queue is examined. If it is a mouse event, it is popped. The change in position for both events is combined by adding changes in the X direction and the Y direction. This effectively is vector addition. It is even simpler if the device, such as a dial, has only one dimension. When two events are combined, the queue checks again to see if the next event is of the same type and another merge can take place. Once the queue is empty or the next event is of a different type, the combined events are processed as a single event.

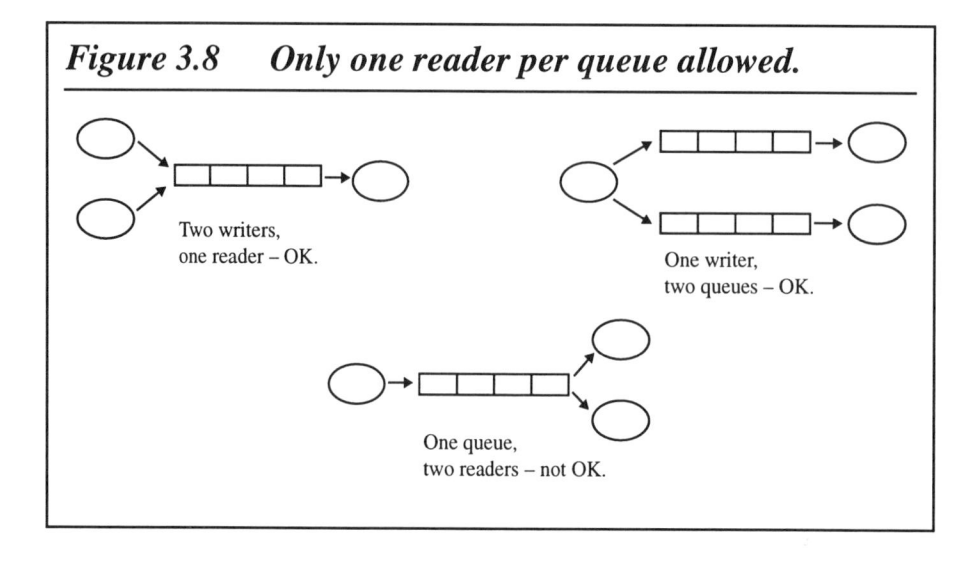

Figure 3.8   Only one reader per queue allowed.

Two writers,
one reader – OK.

One writer,
two queues – OK.

One queue,
two readers – not OK.

With this technique, many intermediate steps may be skipped. If the processor is fast enough to keep up with the events, there will never be more than one event in the queue and no events combined. As the processor falls further behind, the queue grows longer and more events are combined on each pass. If the user moves the mouse quickly, he will see the mouse redrawn only in a few positions. If the mouse is moved more slowly, it is redrawn at each intermediate step. This behavior should be more acceptable to the user than allowing the mouse displayed on the screen to lag the movement of the user's hand. Similarly, if a dial updates a numeric display, then the faster the user turns the dial, the larger the numeric value change each time the display is updated.

Take care if the intermediate values are recorded. Consider a drawing tool that lets the user draw a freehand curve with a mouse or other pointer. A line is drawn that joins all the points on the screen visited by the mouse. The optimization just described might result in fewer longer lines between the points as a result of the merge, rather than a greater number of short lines.

In order to implement this technique, it has to be possible to look at the top element of the queue without removing it. This allows the main event-handling loop to pop the element later if it is an event of a different type. Some RTOSs allow the program to examine the top element of the queue without removing it. If yours does not, you can add this functionality by adding a layer above the RTOS-supplied queues. This layer pops an element and stores it when a read is requested. If the element is subsequently popped, it returns the stored element as the top of the queue. This code does not need to be made reentrant because it would be called only from the task that is reading the queue, and as I have already discussed, there should be only one reader.

3.8 Directing Event Traffic

Once an event is popped off the queue, the controller must decide what the event really means in the current context. Some events have a single purpose. Other events, such as the UP and DOWN keys in the temperature/pressure/flow example, perform different actions depending on the context. To discuss this issue in detail, I will first examine more code from the example. I will use the architecture in Figure 3.4, in which the model, controller, view, and physical output layer all live in the one task. The model data is stored in the following data structures.

```
typedef struct {
    int temperature;
    int pressure;
    int flow;
} Settings;
```

```
static Settings GS_settings;
typedef struct {
    Boolean highTemperature;
    Boolean highPressure;
} Alarms;
static Alarms GS_alarms;
```

The alarms are updated simply by the handleAlarm() function, which is called from the infinite loop of the mvc task, and which you have already seen.

```
void handleAlarm(AlarmStruct *alarmPtr)
{
    switch (alarmPtr->type)
    {
    case HIGH_TEMPERATURE:
        /* Turn alarm on or off as appropriate */
        GS_alarms.highTemperature = alarmPtr->value;
        break;
    case HIGH_PRESSURE:
        /* Turn alarm on or off as appropriate */
        GS_alarms.highPressure = alarmPtr->value;
        break;
    default:
        assert(FALSE);
    }
}
```

This function makes simple changes to the model, which are then copied to the view by the refreshAlarms() function. In the DOS example, this function prints the strings on the screen. A real embedded system may turn lights on instead.

```
void refreshAlarms(void)
{
    if (GS_alarms.highTemperature)
    {
        gotoxy(40,15);
        puts(" HIGH TEMPERATURE");
    }
    else
    {
        /* Clear the alarm if it was displayed */
        gotoxy(40,15);
        puts("                   ");
    }
```

```
    if (GS_alarms.highPressure)
    {
        gotoxy(40,17);
        puts(" HIGH PRESSURE");
    }
    else
    {
        /* Clear the alarm if it was displayed */
        gotoxy(40,17);
        puts("                    ");
    }
}
```

3.8.1 The Focus

Manipulation of the model for temperature, pressure, and flow settings is not so simple. Pressing the key associated with one of these settings causes it to be displayed. You must keep track of the one that is displayed, which is called the *focus*. The concept of focus is well established on desktop user interfaces where the keyboard focus is on a particular window depending on the mouse position. A focus is useful when some piece of hardware can be directed to control a number of different data items. Such a multiplexed control uses the focus to decide which of the data items to control. Changing the focus redirects the events generated by the multiplexed control or changes the interpretation of those events. In this example, the value of the focus is stored in GS_focusValuePtr.

```
/*
The GS_focusValuePtr points to the value in the model that is being
modified.
*/
static int *GS_focusValuePtr = NULL;
```

When a key event for one of the settings occurs, the focus is adjusted by setting GS_focusValuePtr to point at one of the three values in GS_settings. If the key is an arrow key, the value in the model is adjusted via the pointer.

```
void handleKey(KeyValue key)
{
    switch (key)
    {
    case TEMP_KEY:
        GS_focusValuePtr = &GS_settings.temperature;
        break;
```

```
        case PRES_KEY:
            GS_focusValuePtr = &GS_settings.pressure;
            break;
        case FLOW_KEY:
            GS_focusValuePtr = &GS_settings.flow;
            break;
        case UP_KEY:
            /* Increment the focus value if there is one */
            if (GS_focusValuePtr != NULL)
            {
                (*GS_focusValuePtr) ++;
                if ((*GS_focusValuePtr) > 100)
                {
                    *GS_focusValuePtr = 100;
                }
            }
            break;
        case DOWN_KEY:
            /* Decrement the focus value */
            if (GS_focusValuePtr != NULL)
            {
                (*GS_focusValuePtr) --;
                if ((*GS_focusValuePtr) < 0)
                {
                    *GS_focusValuePtr = 0;
                }
            }
            break;
        default:
            assert(FALSE);
    }
}
```

This mechanism of using a focus is implemented with a pointer to the integer to be changed. A more realistic case would use a pointer to a structure, and this structure would contain information such as maximum and minimum limits to restrict the amount of change the user can make to the value.

A slightly different implementation is needed if the data is of different types, say, an integer and a floating-point number or a sequence of strings used in a text menu. However, the same principle can be applied by using an enumerated type to identify one of several types of update that can be performed. In some cases, the data structure that represents the view, rather than the model, can be used. In C++, the focus is often represented by a pointer to an object.

Figure 3.9 shows that the focus of the arrow keys has a choice among the three settings available.

A focus is most useful when a piece of input hardware is shared among several parts of the model.

3.8.2 Callbacks

A callback is a function called in response to an event. Its name stems from a module, object, or subsystem wanting to say: "If event X happens, then call me back." The function to be called is then recorded in some way so that it can be invoked when X occurs. If either the number of focuses or the number of enumerated values used by the focus are large, it can become difficult to manage and callbacks may be a good alternative. If the choice is among several actions rather than among several pieces of data, you should consider applying callbacks.

This mechanism matches the requirements of buttons well. Each button is associated with a callback function; that is, the action to be performed when the button is pressed. In the `handleKey()` function already discussed, a `switch` statement is used to distinguish among the buttons. Unfortunately, the `switch` statement is static; it cannot be changed at run time. To examine how callbacks can be used on a device in which the buttons change meanings, I will use the example of a simple digital watch. This digital watch in Figure 3.10 has three buttons. It has two modes: time-of-day mode and stopwatch mode. Depending on the mode, the buttons change meaning. In typical digital watch style, I have written each of the meanings on each of the buttons, and the user has to figure out on the fly which one applies. Most digital watches use a small icon in the LCD to distinguish modes. Because I wrote this example in text mode in DOS, the mode is displayed as an explicit string below the watch face. Note also that the time of day can be adjusted in the normal time-of-day mode. There is no special mode for updating the current time such as you would find on most real digital watches. Those simplifications aside, this is a suitable case for callbacks because the buttons change their associated actions depending on the mode and on whether the stopwatch is running. This watch displays hours and minutes in time-of-day mode and minutes and seconds in stopwatch mode.

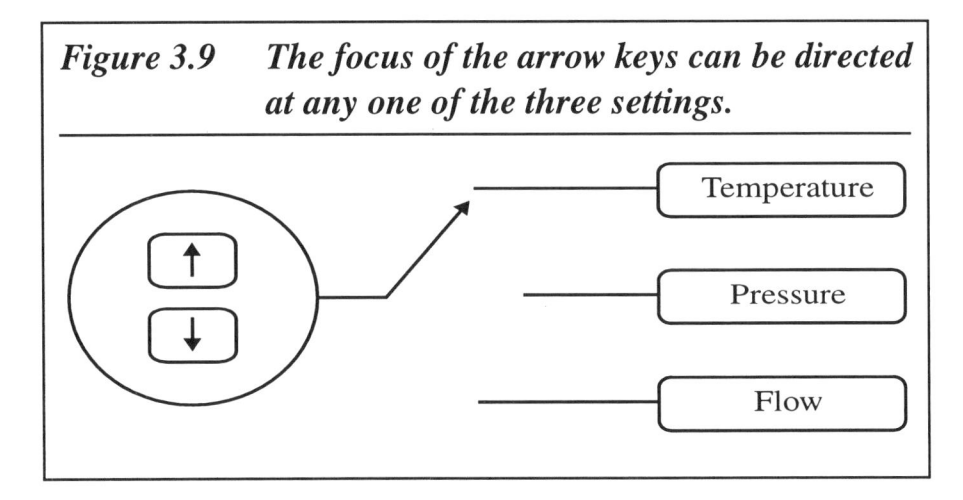

Figure 3.9 The focus of the arrow keys can be directed at any one of the three settings.

If each button has a function associated with it and those functions are stored in an array of function pointers, then you can change them at run time to alter the behavior of the buttons. The following definitions allow us to use such an array.

```
typedef enum  { BOTTOM_LEFT_KEY, TOP_RIGHT_KEY,
                BOTTOM_RIGHT_KEY, NO_KEY} KeyValue;

#define NUMBER_OF_KEYS 3

typedef void (*CallbackFn)(void);

/*
This is the array that will hold the callback for each key
*/
static CallbackFn GS_keyCallbacks[NUMBER_OF_KEYS];
```

The keys are polled in the same fashion as the previous example, but the packets on the queue are a little simpler because there are no alarms to pass. The timeout message is replaced with a one-second timer used to drive the ticking of the watch. The new packet is as follows.

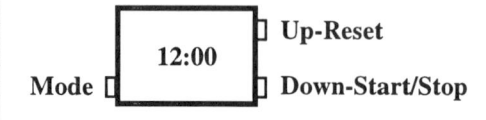

Figure 3.10   The watch example.

```
                  ☐ Up-Reset
         12:00
Mode ☐            ☐ Down-Start/Stop
```

        Mode = Time-Of-Day

The left-hand button changes between time-of-day mode and stop-watch mode. The two buttons on the right have different meanings, depending on the mode. In stop-watch mode, the lower-right button can mean start or stop, depending on whether the stop-watch is running.

Scheduling and Managing User Events — 51

```
typedef enum { KEY_PACKET, ONE_SECOND_TIMER_PACKET } PacketType;

/*
This packet does not require a union because there is only
one packet with an associated value.
*/
typedef struct {
    PacketType type;
    KeyValue keyValue;
} QPacket;
```

The data being manipulated is the time of day and the current time on the stopwatch. Whether the time of day or the current stopwatch time is displayed is in the domain of the view.

```
/* Model Data */
typedef struct {
    int hrs;
    int mins;
    int secs;
} TimeOfDay;
typedef struct {
    Boolean running;
    int mins;
    int secs;
} StopWatchTimer;
static TimeOfDay GS_timeOfDay;
static StopWatchTimer GS_stopWatch;
/* View Data */
typedef enum { TIME_OF_DAY_MODE, STOP_WATCH_MODE } DisplayMode;
static DisplayMode GS_displayMode = TIME_OF_DAY_MODE;
```

Even though the seconds of the time of day are stored, they are not displayed on the watch face.

Now that the data structures are in place, I can initialize the callbacks as follows.

```
GS_keyCallbacks[BOTTOM_LEFT_KEY] = setModeToStopWatch;
GS_keyCallbacks[TOP_RIGHT_KEY] = incrementTimeOfDay;
GS_keyCallbacks[BOTTOM_RIGHT_KEY] = decrementTimeOfDay;
```

The entries in the array are assigned to functions that I will now define. The two mode-change functions, setModeToTimeOfDay() and setModeToStopWatch(), redefine the buttons on the right-hand side of the digital watch to functions appropriate to the new mode. The bottom-right button also changes roles, depending on whether the stopwatch is running.

```
void setModeToTimeOfDay(void)
{
    GS_keyCallbacks[BOTTOM_LEFT_KEY] = setModeToStopWatch;
    GS_keyCallbacks[TOP_RIGHT_KEY] = incrementTimeOfDay;
    GS_keyCallbacks[BOTTOM_RIGHT_KEY] = decrementTimeOfDay;
    GS_displayMode = TIME_OF_DAY_MODE;
}
void setModeToStopWatch(void)
{
    GS_keyCallbacks[BOTTOM_LEFT_KEY] = setModeToTimeOfDay;
    /*
    Bottom left means start or stop depending on whether we are running.
    */
    if (GS_stopWatch.running == TRUE)
    {
        GS_keyCallbacks[BOTTOM_RIGHT_KEY] = stopStopWatch;
    }
    else
    {
        GS_keyCallbacks[BOTTOM_RIGHT_KEY] = startStopWatch;
    }

    GS_keyCallbacks[TOP_RIGHT_KEY] = resetStopWatch;

    GS_displayMode = STOP_WATCH_MODE;
}
```

The following two functions perform the tasks associated with the two right-hand buttons while in time-of-day mode.

```
void incrementTimeOfDay(void)
{
    GS_timeOfDay.mins ++;
    if ( GS_timeOfDay.mins == 60 )
    {
        GS_timeOfDay.mins = 0;
        GS_timeOfDay.hrs ++;
        if ( GS_timeOfDay.hrs == 24 )
        {
            GS_timeOfDay.hrs = 0;
        }
    }
}
```

```
void decrementTimeOfDay(void)
{
    GS_timeOfDay.mins --;
    if ( GS_timeOfDay.mins == -1 )
    {
        GS_timeOfDay.mins = 59;
        GS_timeOfDay.hrs --;
        if ( GS_timeOfDay.hrs == -1 )
        {
            GS_timeOfDay.hrs = 23;
        }
    }
}
```

The three stopwatch functions are associated with the two right-hand buttons while in stopwatch mode. The bottom-right button is associated with `startStopWatch()` or `stopStopWatch()`, depending on whether the stopwatch is running.

```
void startStopWatch(void)
{
    GS_stopWatch.running = TRUE;
    GS_keyCallbacks[BOTTOM_RIGHT_KEY] = stopStopWatch;
}

void stopStopWatch(void)
{
    GS_stopWatch.running = FALSE;
    GS_keyCallbacks[BOTTOM_RIGHT_KEY] = startStopWatch;
}
void resetStopWatch(void)
{
    GS_stopWatch.mins = 0;
    GS_stopWatch.secs = 0;
}
```

The infinite loop of the `mvc` task is of the same form as the temperature/pressure/flow example, but the `handleKey()` function has changed dramatically. Instead of having a large switch statement to handle all possibilities, it now has only to activate the appropriate callback.

```
void handleKey(KeyValue key)
{
    GS_keyCallbacks[key]();
}
```

This function is simple because all buttons use callbacks. In some cases, you may use a mixture involving a `switch` statement to cover the keys or events whose meanings never change and the callback mechanism for the remainder.

The remainder of the code for this example manages the timing for the seconds ticking by and for refreshing the display, neither of which relates to the callback mechanism directly, so I will not investigate it further. You will see more of the callback mechanism when you look at the code for buttons on graphical displays.

One word of caution about using callbacks: They make the flow of control harder to predict by reading the code. In the example just given, any meaning could be attributed to a button at run time, and there is no way in C to limit that flexibility. Programmers who use Smalltalk see this as a natural form of message passing. However, it makes tracking defects difficult. If you have a set of callback functions for mechanical buttons and a completely different set of callback functions for graphical buttons, you may want to guarantee a compiler warning if you assign a callback of one type when a different one is expected. This is possible if you give the two types of callback different signatures, as in the following example.

```
typedef void (*GraphicalCallbackFn)(void);
static GraphicalCallbackFn GS_graphCallbacks[NUM_GRAPH_BUTTONS];
typedef void (*MechanicalCallbackFn)(int mechanicalButtonIndex);
/*
This is the array that will hold the callback for each
mechanical button
*/
static MechanicalCallbackFn GS_mechCallbacks[NUM_MECH_BUTTONS];
void someMechanicalCallbackFunction(int mechanicalButtonIndex);
void someGraphicalCallbackFunction(void);
void setCallback(void)
{
    /* This line is OK */
    GS_mechCallbacks[0] = someMechanicalCallbackFunction;
    /* This will generate a compiler warning */
    GS_mechCallbacks[0] = someGraphicalCallbackFunction;
}
```

The second line of the function generates a warning because the assignment is expecting a function that takes an integer as an argument. This does not save you from assigning one of the mechanical button callbacks to the wrong button, but at least it avoids some more obvious errors. In Chapter 7 when I discuss C++, you will see callback mechanisms that are much safer.

Chapter 4

Finite-State Machines and Table-Driven Software

Finite-state machines (FSMs) are frequently used as tools for modeling many systems and provide a means of controlling the flow of a program. They are also easy to represent in diagram form, which has obvious advantages during design and documentation phases.

FSMs are a special case of table-driven code. Table-driven code controls the flow of events using data structures to dictate the order of function calls rather than using explicit calls within a function. It would not make sense to structure all of your code in this way, but in a few areas it can be very effective.

4.1 The FSM as a Poor Man's Real-Time Operating System

The FSMs described in this chapter are not used to schedule pseudo-parallel threads of execution. In simple embedded systems, you may use an FSM to manage the flow of control in order to allocate CPU cycles between a set of logically distinct threads of control. While in a given state, the function associated with that state is executed and that function has control of the processor until it returns. On completion of the function associated with one state, the state machine progresses to the next state and runs the function associated with this new state. This loop continues with each state, giving each function a chance to run. Such an implementation is a poor man's RTOS. Using an FSM in such a way works on small systems but does not scale well, and it makes it difficult to implement any sort of priority scheme. FSMs are completely unsuitable if some of the threads of control have hard real-time deadlines and others do not. In such a scenario, an RTOS is required. I describe this type of FSM to distinguish it from the FSMs I am concerned with in this chapter. The FSMs implemented here are used purely to manage the sequence of inputs from the user and have no effect on scheduling CPU time.

4.2 Drawing the FSM

An FSM consists of a set of states and a set of transitions between those states. In any given state, there are a set of legal input values. Each value causes one transition to occur.

The simple example in Figure 4.1 shows the three possible states of a door. The door may be Open, Closed, or Locked. In order to Lock the door, it must be Closed. When the door is unlocked from the Locked state, it becomes Closed again. There are no transitions from the Open state to the Locked state. This means that no single input could cause this transition. If the door is Open and the input Lock is applied, no transition occurs. This input is considered illegal. As the machine's implementor, it is up to you to decide if illegal actions are ignored, flagged to the user as an error, or cause the program to halt. The source of the input will affect your decision. Sometimes, the input is the keys the user pressed, in which case, an illegal transition simply means that the user pressed a key with no associated action. In other cases, the input is filtered before it reaches the FSM, and an illegal input may represent a bug in the filter.

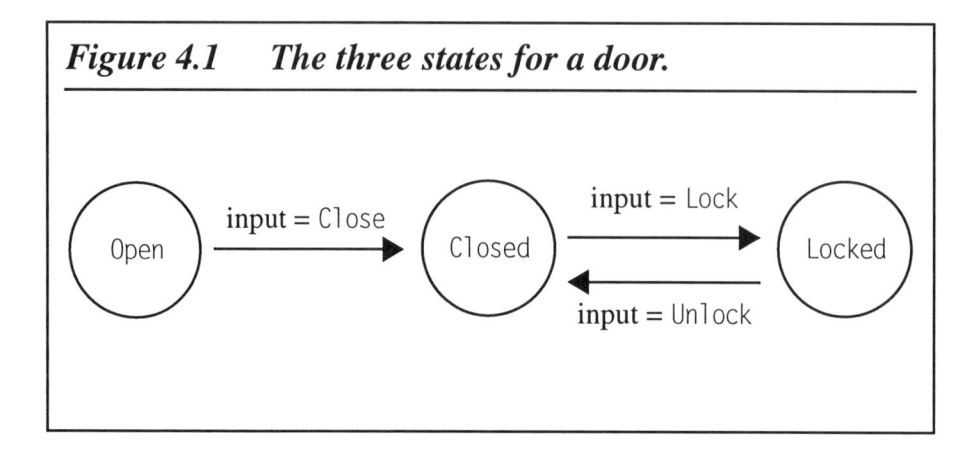

Figure 4.1   The three states for a door.

You should always be aware of how visible the FSMs are to the user. Some FSMs are used to manage processes of which the user is not aware. In other cases, I want the current state of a machine, or machines, to be visible to the user. For example, the element currently highlighted in a menu represents a certain state. Sometimes, the state itself is not visible, but the transitions are visible to the user. Each transition may be associated with an action function, and that function may interact with the user, making the user aware that a transition took place.

I will repeat the door example here with a slightly different syntax to show a function call associated with each input. Let us assume that the door FSM must generate the appropriate noises when the user makes a change. Those noises are generated by the functions Slam() and Click(). Figure 4.2 shows this slightly more capable FSM.

For the purists, an FSM that has an action associated with each transition is a Mealy FSM. Alternatively, you may associate the actions with each state. Such an FSM is called a Moore FSM (Hendricksen, 1989). I will not discuss the Moore FSM variety further.

4.3 Why User Interface Code Needs FSMs

In a user interface, the use of FSMs is more common than in many other areas of programming. One reason for this is that the FSM is used to store state across numerous events, or inputs. The current state gives a context to the next input. This context dictates the meaning of the input. In many cases, the response to a user input depends on the history of previous inputs. In other cases, the response depends on the history of previous inputs to a particular area of the interface but is completely independent of inputs to other areas. A filter on all the inputs can decide which inputs drive each FSM.

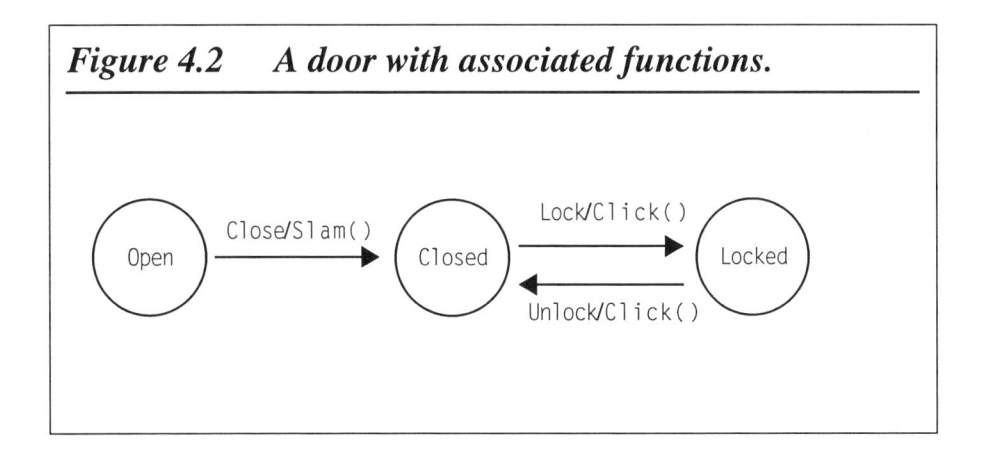

Figure 4.2   A door with associated functions.

Much of the time, the user is aware of this state information. He realizes that pressing the Quit button while in a submenu has a different effect than pressing the Quit button while at the top-level menu.

4.3.1 States and Modes in the User Model

Where separate states represent separate modes of operation, it is important that the mode be apparent to the user. On a word processor, any time I find myself typing in over-type mode when I thought I was in insert mode, I curse the fact that I did not spot the mode I was using. (Modes have a bad reputation and are considered contributors to unusable interfaces. It is more often because the mode is not visible or obvious to the user rather than the mode itself that causes the difficulty.) Consider a drawing tool in which the modes are Draw_Line, Draw_Box, and Draw_Arc. If the user cannot tell the mode from the appearance of the interface, he may drag the mouse and only then realize he is drawing a box when he intended to draw a line. The user is forced to undo the action just performed and then change to the correct mode. If the cursor is in the shape of a small box, however, the user is unlikely to make that mistake. This is an example of making the FSM (with transitions between the Draw_Line, Draw_Box, and Draw_Arc states) more visible to make navigation easier for the user.

It may be tempting to use an FSM to implement an Undo facility. This is not as simple as it seems. If an Undo key retraces the last transition performed, there should also be a transition function that can reverse the effect of the last transition. It is also important to check if the event could have been interpreted on a different level than the level at which the FSM has been implemented. As long as this warning has been observed, it can be useful to record every state visited in order to provide a multistep Undo feature.

4.4 Limits of the FSM

An FSM has no memory apart from the current state. A decision can be made based on the current state, because there is a different transition for each of the possible last states. However, it is not possible to make a decision based on the last time this state was visited. If you want to build a machine that allows the program to visit each state once, at most, data would have to be maintained outside the FSM to record the states that have been visited. More powerful machines can be built at a cost. You may add new rules to the machine, such as a valid/invalid flag that can be toggled for each state.

Depending on actions outside the machine, some states can change their behavior. This may be useful. However, once you go beyond the basic elements, the machine is more difficult to represent in a simple diagram that other engineers can understand. If the FSM has been enhanced, the reader will have to learn those enhancements before he can understand the diagrammatic representation. For FSMs that are visible to the user, this is an important trade-off, because the state diagram can often be used to document the behavior of the interface. Unfortunately, many real-world programs have characteristics that are too complex to model with a simple FSM. More complex mechanisms must be used. The menu example later in this chapter shows such a system.

4.5 How Many States? How Many FSMs?

Having several states per interface allows the limited number of inputs available to the user more meanings than would be otherwise possible. Each input can have a different meaning in each state, allowing the number of meanings to be the number of possible inputs multiplied by the number of states. The maximum number of possibilities, however, are rarely used. These meanings, in FSM terminology, are transitions.

Do not assume that the interface should be seen as one large FSM. It is often more useful from the user's and programmer's point of view to model the interface as a collection of FSMs. One FSM may model the possible states of a button (which is not as simple as you might think). Several instances of that FSM are required if there are many buttons. Another FSM may model the different states of the whole system: Booting, Running, and Self_Test_Mode. There would be only one such FSM per system.

Attempting to combine many FSMs into one has some attractions. The specification of the interface could be represented in one large diagram. Testing could be performed according to how many states have been visited, and one of the test completion goals would be that all states must have been reached during the test process.

Unfortunately, the disadvantages far outweigh the advantages. One large FSM that combines all possible states for an interface would have more states than the sum of the smaller FSMs. Consider two FSMs; call them A and B. The number of states in an FSM that combines all the possibilities is the product of the number of states in A and the number of states in B. Not only have the number of states increased, but the meaning of each state has grown more complex. When the FSMs are decomposed, there may be a state that represents Menu_Is_Selected and another state in the other FSM that represents Key_Is_Depressed. In the combined FSM, there is a state called Menu_Is_Selected_And_Key_Is_Depressed.

Although few programmers might attempt to combine an entire complex interface into one FSM, it is tempting to combine related FSMs. If you have a large and complex FSM, consider decomposing it into several simpler FSMs. If you have several small FSMs that depend on each other's behavior, you may have a candidate for combining them into one.

4.6 Use of Constant, Static, Global, and Automatic Data

Two types of information must be stored to implement an FSM: the table itself, which is information that does not change, and the current state, which changes throughout the life of the program. There may be many copies of the current state if there are many instances of the FSM in use within the program.

The table that represents the states and transitions is declared `const`. The `const` declaration instructs the compiler to place the table in the code segment and may, therefore, be placed in ROM.

Much of the state information must be maintained between events. After an event is handled, the program and the stack return to the point where they read the event queue for the next input. All the functions that have been called to handle the event have returned and unwound the stack by the time the next event can be handled. The state, therefore, has to be held in data structures that are not automatic variables of any of those functions. Note that automatic means local variables that are on the stack. Local variables can also be constant, in which case they can be accessed from within the function but not changed, or they can be `static`, in which case they can be accessed from within the function, but they hold their value between invocations. The value is, therefore, not stored on the stack.

This allows two options for storing the current state. It may be stored as a `static` variable. The variable has a life that begins when the program starts running and continues until the program terminates. These variables may have local function scope, file scope, or global scope. Alternatively, the space may be allocated dynamically on the heap by `malloc()`.

4.7 Example of an FSM for a Toggle Button on a GUI

This example implements a toggle-button on a graphical interface. The button latches itself Down when pressed once. If the user presses it again, it latches itself Up. The text visible in the button describes the attribute being toggled. For example, the word "Silence" may indicate that while in the Down state, this button is silencing some speaker or alarm buzzer. This example was devised for a touch screen, so I assume that the user presses the graphically depicted button with his finger, although he could as easily be using a mouse to select it. An FSM could also control a mechanical button with an associated LED that reflects its state.

4.7.1 The Four States: Up, Down, Pending_Up, and Pending_Down

Although the user is aware of the Up and Down states, he is less conscious of the state of the button while his finger is on it. These states are necessary because it is possible for the finger to slide off to one side without completing the operation. The rule for the button is that the finger must be removed from the screen while the button is selected to complete the operation. This avoids accidentally selecting two neighboring buttons in a single operation. After an aborted selection, the button must know what state it should revert to. If you did not have the intermediate states, the state would change from Up to Down when the finger pressed on it. It would revert to the Up state if the user slid his finger off the button. This would turn the toggle on and then off, which may not leave the system in the same state as not turning it on at all. So, if the user slides his finger off before a release, you will treat this as not selecting the button.

The transition diagram in Figure 4.3 shows the possible states and the transitions between them. The doNothing() function acts as its title suggests. Such functions are useful when there are slots in the transition table that you prefer to leave empty. Having an empty function avoids having to check if the function pointer is null each time the transition occurs.

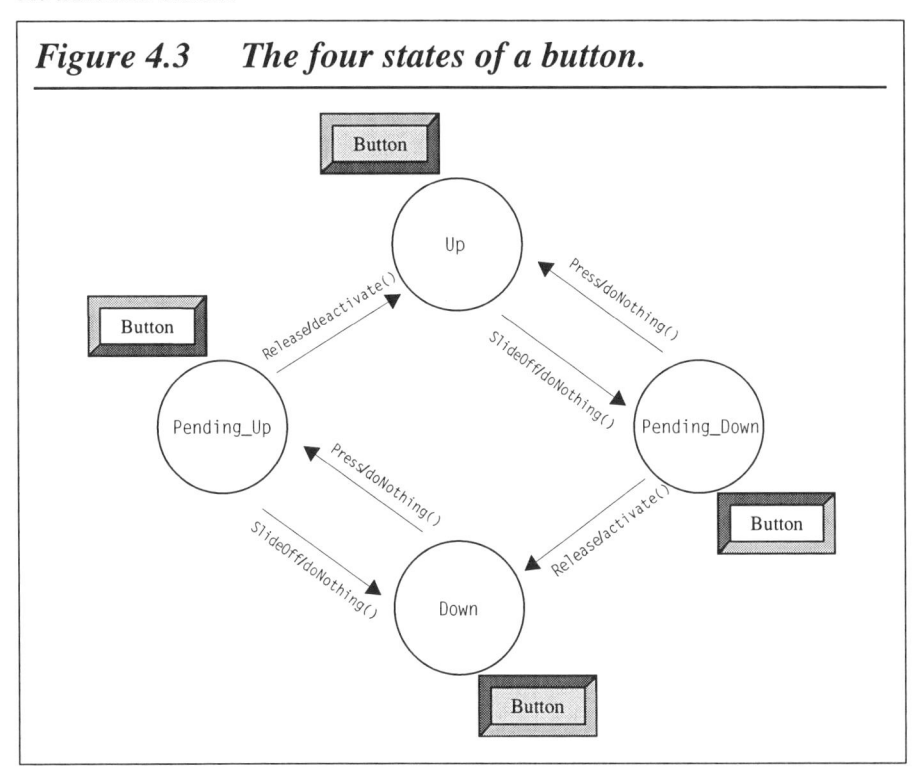

Figure 4.3 **The four states of a button.**

The button images are shown next to each state. Drawing the image could be performed by copying prestored bitmaps. In the example code, the image is drawn by constructing a number of lines, rectangles, and polygons. The drawing method is not relevant to the code to manage the FSM. The button's appearance reflects the state, so it is straightforward for the user to see each state. The 3-D effect is useful to distinguish touch-sensitive buttons from other labels on the screen that are not touch-sensitive. At this point, I suggest that you run the `fsm` program supplied on the companion disk. It allows you to view the button in each of the described states by selecting it with the mouse.

The possible inputs are `Press`, `SlideOff`, and `Release`. `SlideOff` occurs when the user moves his finger out of the area of this button. Whether he enters the area of another button is immaterial to this FSM, because this FSM is concerned with only one button, not with the entire set of buttons on the screen.

These inputs cannot be read directly. As each finger position is read from the display, it is parsed, or filtered, to detect any higher level events it represents. The button object may have functionality apart from the FSM that indicates whether the button area contains the point selected by the finger. A function can then process each position, establish exactly which buttons are affected by the latest event, and generate the appropriate inputs. Such processing, although interesting, is not the subject of this chapter; see Chapter 5 for further details.

By happy coincidence, this button works nicely if you treat a finger sliding into the area of the button exactly the same as a `Press`. This avoids having to create another input value. If the user slides his finger over the display, entering and leaving several `Up` buttons, the only one that transitions into the `Down` state is the one in the `Pending_Down` state when the finger is lifted from the screen.

4.7.2 The Interaction Between the FSMs

In the example, each instance of a `Button` contains an instance of `ButtonState`, which is all the information you need to store to have a unique FSM. Each individual FSM obeys the same transition rules. However, the state of one FSM has no influence on the state of any other.

In the example, the transitions are stored as `const` data. Thus, they cannot be altered while running. If the FSM did change at run time, it would not be possible to share the transition table among many instances of the FSM, because changes to the table that implements the FSM would influence all instances.

Many of the type names in the example restrict the function `ButtonProcessFSM()` to handle only buttons. The basic algorithm could be implemented in a more general way and then used for all FSMs in your system. If data, such as a pointer to the `Button` structure, needs to be passed, it can be passed as a `void` pointer. All the enumerated values would be passed in as integers. This involves casting, which is less type-safe. In return, you can reuse the same function. Despite the code reuse, I do not usually create a more general FSM-processing function. Most real-world cases involve some exceptions. For example, you may wish to change the rules that decide when transitions are legal. One enhancement is allowing the action function to return a Boolean value and allowing the transition only if the action function returns TRUE. In other cases, you may wish to transition to a certain state if an illegal input is received.

If there are many transitions, you may want to sort them and allow a binary search for the matching input, or you may wish to sort them according to frequency of use, to allow the most frequent transitions to be found soonest in a linear search.

Amid all the special-case processing, the few lines of code that actually call the action function and change the state represent a small percentage, so the code reuse advantage is minor.

4.7.3 The Implementation

Each transition is represented by a constant instance of the ButtonTransition structure. The structure is shown here, along with a set of type definitions required to define the transition.

```
typedef enum  {NullInput=0, Press, Release, SlideOff} ButtonInput;
typedef enum  { UP, PENDING_DOWN, DOWN, PENDING_UP } ButtonState;
#define NUMBER_OF_STATES 4
/*
This typedef defines a pointer to an action function.
The arguments pass all of the information associated with this
transition in case the action function based on that information.
*/
typedef void (*ButtonFunction)(Button *buttonPtr, ButtonInput input,
            ButtonState nextState);
/*
The transition consists of the input, the next state, and
the action function.
*/
typedef struct  {
    ButtonInput input;
    ButtonState nextState;
    ButtonFunction transitionFn;
}ButtonTransition;
```

Each state has a list of such transitions. Each transition is checked for an input matching the user input. When a match is found, transitionFn is called and the state is assigned to nextState.

The following assignments represent the full set of transitions. Note that the final entry in each array starts with a NullInput value. This is a terminator and the rest of the last line is not used.

```
const ButtonTransition UPtransitions[] =
{
/*      INPUT           NEXT STATE      ACTION */
        { Press,        PENDING_DOWN,   doNothing },
        { NullInput,    UP,             doNothing }
};
const ButtonTransition PENDING_DOWNtransitions[] =
{
/*      INPUT           NEXT STATE      ACTION */
        { Release,      DOWN,           activate },
        { SlideOff,     UP,             doNothing },
        { NullInput,    UP,             doNothing }
};
const ButtonTransition DOWNtransitions[] =
{
/*      INPUT           NEXT STATE      ACTION */
        { Press,        PENDING_UP,     doNothing },
        { NullInput,    UP,             doNothing }
};
const ButtonTransition PENDING_UPtransitions[] =
{
/*      INPUT           NEXT STATE      ACTION */
        { Release,      UP,             deactivate },
        { SlideOff,     DOWN,           doNothing },
        { NullInput,    UP,             doNothing }
};
const ButtonTransition *GS_buttonFSM[NUMBER_OF_STATES] =
{
    UPtransitions,
    PENDING_DOWNtransitions,
    DOWNtransitions,
    PENDING_UPtransitions
};
```

The button itself is represented as a structure. Although one FSM can be used by all buttons, I must create one Button structure for each button.

```
struct buttonStruct;
typedef struct buttonStruct Button;
struct buttonStruct
{
    int x;
    int y;
    char *text;
    ButtonState state;
    void (*actionFnDOWN)(Button *buttonPtr);
    void (*actionFnUP)(Button *buttonPtr);
};
```

The state field in this structure represents the current state of this button within the FSM. The two function pointer fields are used to call particular functions once the button has reached the Up or the Down states. There is no function for the

Pending_Down or Pending_Up states, because no permanent transition has taken place. Note that these two functions will be different for each button, because each button has a different job to do.

The buttonProcessFSM() function takes an input and applies it to the Button that is pointed to by the first argument. Note that only the set of transitions associated with the current state must be checked. Other transitions are ignored.

```
void buttonProcessFSM(Button * buttonPtr, ButtonInput input)
{
    int i;
    /*
    First loop through the FSM to find the appropriate transition
    */
    for (i=0; GS_buttonFSM[buttonPtr->state][i].input != input; i++)
    {
        if (GS_buttonFSM[buttonPtr->state][i].input == NullInput)
        {
            /*
                The end of the list has been reached and no input
                matched the input
            */
            assert(FALSE);
        }
    }
    /*
    Some transition is going to take place so set flag to indicate
    that the buttons need to be redrawn.
    */
    GS_changeHappened = TRUE;
    /* Now i indexes the appropriate transition */
    /* Call the action function, passing the button, the input and
       the nextState */
    GS_buttonFSM[buttonPtr->state][i].transitionFn(buttonPtr, input,
                GS_buttonFSM[buttonPtr->state][i].nextState);
    /* Change the state */
    buttonPtr->state = GS_buttonFSM[buttonPtr->state][i].nextState;
}
```

The three transition functions follow. Some transitions perform no action. When the button is placed in the Down state or the Up state, one of the button actions takes place. The actionFnDOWN and actionFnUP pointers may point to any function that takes no arguments and returns void. These are the callbacks for this button. These functions perform the application-level work in response to the user's action on the button.

```
void doNothing(Button *buttonPtr, ButtonInput input,
               ButtonState nextState)
{
    /* This function simply returns */
}
void activate(Button *buttonPtr, ButtonInput input,
              ButtonState nextState)
{
    buttonPtr->actionFnDOWN(buttonPtr);
}
void deactivate(Button *buttonPtr, ButtonInput input,
                ButtonState nextState)
{
    buttonPtr->actionFnUP(buttonPtr);
}
```

Note that two levels of function pointers are traversed here. The first level uses pointers stored in the FSM table. These are the same for all buttons and represent the controller communicating with the model. The second level are functions associated with each button. These may be different for each button, allowing each button to have different application-level behavior. In the example, the FSM's action functions simply call the buttons actionFnUP() or actionFnDOWN(). In a real case, the button's activate() and deactivate() functions may sound a key click, change the cursor's appearance, or add other properties to the button.

4.7.4 A Model-View-Controller Interpretation of this FSM

This FSM is part of the model data. The controller provides a stream of inputs, having filtered the finger positions to interpret them as Press, Release, and SlideOff events. In the demo program, the view is refreshed once the buttonProcessFSM() has returned. The view reads the state from the model (i.e., the state field of the button), decides which color to use to display the button, and decides whether to draw the button in a raised or sunken state.

4.8 Menu for a Small Text Display

This example implements a menu for a small, one-line text display. It demonstrates a more powerful table than an FSM. The menu hierarchy means that I still have a simple graphical method of representing the data in the table. There are no hard and fast rules for which scenarios suit table-driven code and which ones do not. Unlike FSMs, there is no mathematical background to more general tables. One way of thinking about them is as a more general form of the switch statement.

Menus were once a common way to navigate text interfaces on desktop computers. More modern GUIs allow pull-down menus. These menus often give the user access to a dialog to complete the request. Before the age of the GUI, menus often led to submenus, which led to sub-submenus. That's because tick boxes, radio buttons, and other visual gizmos used to construct a request on a GUI were not available on text displays. Many options on a GUI can be shown at the same time, while a hierarchical menu must go through questions one at a time.

Despite the disadvantages, a hierarchical menu is still a necessary tool in embedded systems in which the display device may be a one-line text display or a serial link to a terminal. Even if a full screen is available, the processing ability of the embedded CPU or the amount of development time may limit the interaction to text rather than graphics. The inputs available may be a couple of keys rather than the full keyboard and mouse available on a typical desktop computer.

4.8.1 Human Factor Issues for Menus

The biggest challenge in a menu is to allow the user to say what he wants. The expressive power of a command line has been lost — probably for the better for most novice users. However, arguments may need to come from a question-and-answer session at the lowest level of the menu, making the interaction more time consuming. Limited display space often means that the user cannot see all of his requests at one time. With small displays, there are cases in which the user can see the current option but not the name of the parent menu. If the option facing the user is "Halt job," it is important for the user to remember if the parent he navigated through is "Printer" or "Download." In the implementation presented here, the full path is always visible while navigating. When this is not the case, careful wording of the menu items can reduce ambiguity.

When you design a menu, there is a trade-off between the width and depth of the menu. If each level contains many options, the menu may be only a couple of levels deep. By keeping menus short, you end up with deep menus that are challenging to navigate. A long list in which the items differ in value but not in kind, such as printer types, is acceptable. On the other hand, if the menu itemizes different concepts, such as commands that may be sent to the printer, longer lists are more intimidating.

The magic number seven is often quoted in cognitive psychology texts (Waern, 1989) as the quantity of information chunks that can be retained in short-term memory. That suggests that menus with more than seven options are harder to use than those with fewer than seven. By the time the user looks at option eight, option one has dropped into a less active part of his memory. If keeping the seven rule leads to deeper menus, it is better to break the rule. Despite the psychology texts, experiments show that few levels with many alternatives work better than many levels with few alternatives.

> **Long lists cause navigation difficulty in menus, but deep menus cause more navigation difficulty. Never add depth to reduce width.**

A common navigation aid in menus is to number the items in the menu. This lets the user know how many options he has viewed and whether the menu has wrapped around. However, the number may be misconstrued as a value. If the user sees "3 Pressure" as the third menu item, there is a possibility that he may assume that the value of the pressure is 3. The user may not realize that he must go down one level in the menu to see the actual pressure value.

If you want to be adventurous, you could try inserting earcons (an icon that you can hear) into your menus (Brewster et al., 1996). A sound made in response to a key click in the menu may help prompt the user. This could replace the key click. You would not want a different sound for each menu item, but the sound at a leaf might be different from the sound on reaching a node. The top-level menu could have a unique sound to let you know that you cannot go further up the hierarchy.

4.8.2 The Implementation

The menu is table driven, as was the button in the previous example. The main point here is that, in many cases, adding rules to the table, beyond the basic FSM, allows us to create a more powerful tool.

The hypothetical product is the latest in high-tech running footwear. There is a 30-character display along the sole of the trainer, and the owner may configure the footwear using three buttons marked Up, Down, and Scroll. The colors of the sole, upper, and lace can be changed at run time, if you pardon the pun. The brightness is set as a number — very useful for joggers who venture out at night. The air pressure in the sole is also adjustable via a menu option.

4.8.2.1 The Model

All the information about the trainer is stored in the following structure. The declaration of one instance of the trainer is shown below.

```
typedef enum { RED, GREEN, BLUE, BLACK, WHITE, NUM_COLORS } Color;
typedef struct {
    int pressure;
    int brightness;
    Color soleColor;
    Color upperColor;
    Color laceColor;
} Trainer;
static Trainer GS_trainer    = { 1, 5, RED, RED, BLACK };
```

The table consists of several nodes that form a tree. The tree analogy allows us to use the term *branches* for options at any level and the term *leaf* for a node with no further branches. Each node contains pointers to its parent node, to one child node, and to one sibling, which is the next node on the list at its own level. The parent node points to only one child, and from that child it can reach the other children. This is more convenient than maintaining one pointer per branch in the parent node, because the number of branches will vary.

A pointer to the action function is executed when the node is selected. Only leaves are allowed to have action functions. There is also a string, which is the name of the node and is used for the display.

I use the convention that *down* means further into the menu structure, *up* means back toward the root of the menu, and *next* means the adjacent branch on the current level. Although these definitions of up and down mean that the tree is upside down, it is the usual way to draw menu trees.

```
struct menuNode
{
    /* parent node */
    struct menuNode *upPtr;
    /* default selection of the list of child nodes */
    struct menuNode *downPtr;
    /* The next option in the menu at the current level. */
    struct menuNode *nextPtr;
    /* The function to call if this leaf is selected.
       The actionFn must be NULL if this is not a leaf */
    void (*actionFn)(struct menuNode *nodePtr);
    /* The name of this node for display purposes */
    char * string;
    /* If this node is a leaf, and has a value associated
       with it then it is pointed to by valuePtr.
       It is a pointer to void since the type of the
       value may vary */
    void *valuePtr;
};
typedef struct menuNode MenuNode;
```

A typical element follows.

```
MenuNode mColor={NULL,&mColorSole,&mSettings,NULL,"Color",NULL};
```

The parent is NULL, indicating that this node is at the top level. mColorSole is the first node in the list of children of this node. mSettings is the next item in the list at the current level. The actionFn is set to NULL, because this is not a leaf. The name of this node "Color" is displayed to the user when this node is navigated. The final NULL is the pointer to the value associated with the node, which, again, is unused in this node.

70 — *Front Panel: Designing Software for Embedded User Interfaces*

Figure 4.4 shows the menu tree. This is the conceptual shape of the tree and represents the menu as the user sees it. The pointers stored in each node do not connect in that way, however. The tree in Figure 4.4 requires variable numbers of down pointers, which involves list management. It is simpler to link the siblings, as shown in Figure 4.5, which shows lines representing the actual pointers. The up pointers are not shown. Any node with a NULL down pointer is a leaf and has an associated action function and possibly an associated value.

The table is represented in code by the set of declarations in Listing 4.1. To resolve forward references, all the structures must be declared before they are defined. They are declared `extern` before the set of definitions.

GS_currentNodePtr points at the current node. This gives us a point from which to navigate. It is declared as follows.

```
static MenuNode *GS_currentNodePtr;
```

It is initialized in menuInit() to point to mColor. This allows the menu to be reset at any time by calling menuInit().

The menuDown(), menuUp(), and menuNext() functions perform the menu navigation. menuDown() has the extra responsibility of distinguishing between nodes that have branches and nodes that are leaves. The menuUp() function must realize when it is at the top level and not attempt to go any further. Because all the menus wrap around, there is always a valid nextPtr.

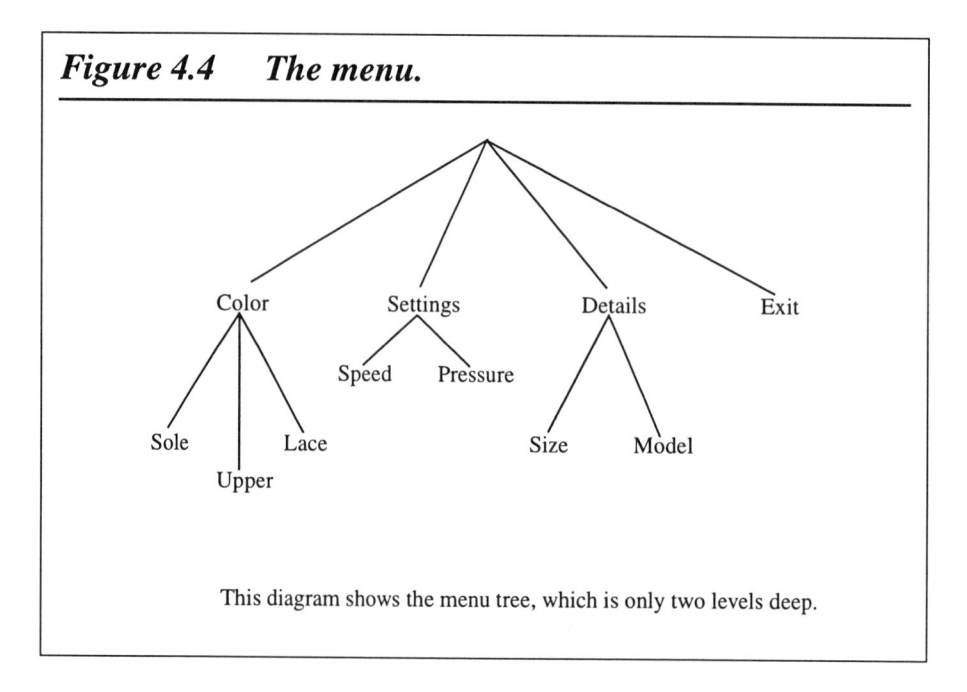

Figure 4.4   The menu.

This diagram shows the menu tree, which is only two levels deep.

```c
void menuNext(void)
{
    GS_currentNodePtr = GS_currentNodePtr->nextPtr;
}
void menuDown(void)
{
    if (GS_currentNodePtr->downPtr)
    {
        GS_currentNodePtr = GS_currentNodePtr->downPtr;
    }
    else
    {
        GS_currentNodePtr->actionFn(GS_currentNodePtr);
    }
}
void menuUp(void)
{
    if (GS_currentNodePtr->upPtr)
    {
        GS_currentNodePtr = GS_currentNodePtr->upPtr;
    }
}
```

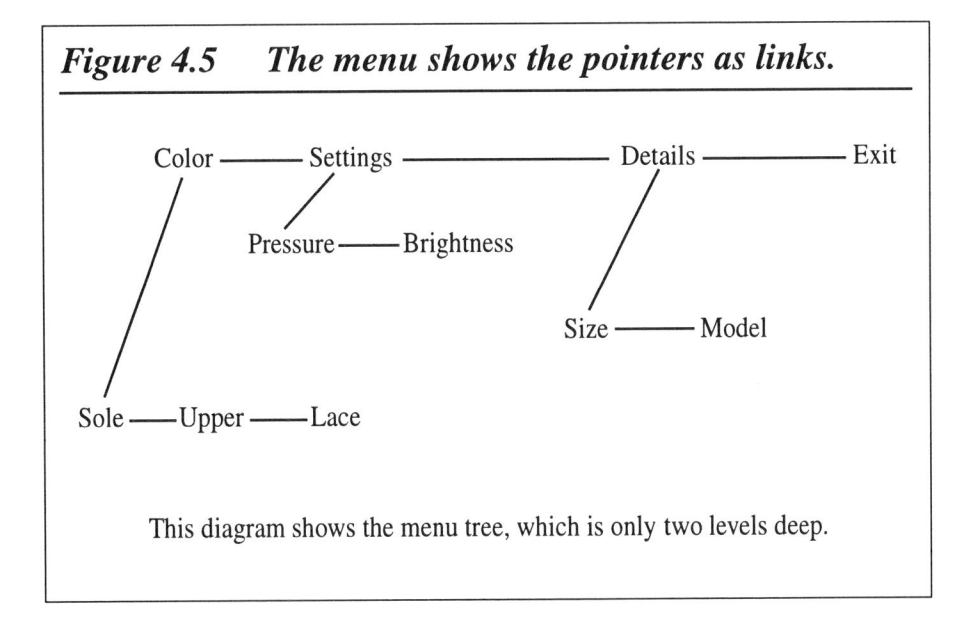

Figure 4.5   *The menu shows the pointers as links.*

This diagram shows the menu tree, which is only two levels deep.

Listing 4.1 The table for a text menu as represented in code by a set of declarations.

```
/*
    UP              DOWN                NEXT            ACTION_FN       STRING          VALUE_PTR
*/
MenuNode mColor={
    NULL,           &mColorSole,        &mSettings,     NULL,           "Color",        NULL );
MenuNode mSettings={
    NULL,           &mSetPressure,      &mDetails,      NULL,           "Settings",     NULL );
MenuNode mDetails={
    NULL,           &mDetailsSize,      &mExit,         NULL,           "Details",      NULL );
MenuNode mExit={
    NULL,           NULL,               &mColor,        exitProgram,    "Exit",         NULL );
/* Expand out the Color sub-menu */
MenuNode mColorSole={
    &mColor,        NULL,               &mColorUpper,   selectColor,    "Sole",         &GS_trainer.soleColor );
MenuNode mColorUpper={
    &mColor,        NULL,               &mColorLace,    selectColor,    "Upper",        &GS_trainer.upperColor );
MenuNode mColorLace={
    &mColor,        NULL,               &mColorSole,    selectColor,    "Lace",         &GS_trainer.laceColor );
/* Expand the Settings sub-menu */
MenuNode mSetPressure={
    &mSettings,     NULL,               &mSetBrightness, selectInt,     "Pressure",     &GS_trainer.pressure );
MenuNode mSetBrightness={
    &mSettings,     NULL,               &mSetPressure,  selectInt,      "Brightness",   &GS_trainer.brightness );
/* Expand the Details sub-menu */
MenuNode mDetailsSize={
    &mDetails,      NULL,               &mDetailsModel, displaySize,    "Size",         NULL );
MenuNode mDetailsModel={
    &mDetails,      NULL,               &mDetails,      displayModel,   "Model",        NULL );
```

Each leaf performs a task, while the internal nodes exist solely to allow navigation. Some leaves manipulate a field of the `GS_trainer` structure. Others just display more information.

Several action functions implement option lists. When the user presses the Down key on an option list, the current value is displayed. The Scroll key can be used to view all possible values. Two asterisks mark the currently selected value, so the user can distinguish it from the other options that he can scroll through.

By pressing the Down key on a menu item, the user may choose the value to apply to the `GS_trainer` structure. For example, pressing the Down key while the display contains `Settings>Pressure = 7` sets the `pressure` field to the value 7. The display changes to `Settings>Pressure = 7 **` to indicate that the new value has been applied.

I need a function to handle changing values. There are two types of values handled in the example program: integers and `Colors`. I will show the code for manipulating integers; the code for `Colors` follows the same form. The first function, `selectInt()`, is an action function pointed at by the menu structures. The second function, `adjustIntProcessKey()`, handles user events once it establishes that he is trying to change an integer. If the events must be directed to the `adjustIntProcessKey()` function rather than one of the navigation functions, `GS_eventHandlingFnPtr` is set to point to the `adjustIntProcessKey()` function. The controller checks this value before performing any navigation. You can think of `selectInt()` as the setup function, while `adjustIntProcessKey()` does most of the real work involved in changing a value.

Because the two functions must communicate with one another, `GS_proposedInt` cannot belong to either one. It, therefore, has file scope.

```
static int GS_proposedInt;
void selectInt(MenuNode *nodePtr)
{
    char intString[MENU_TEXT_LENGTH+1];
    /*
    The event handling function consumes all key events until it is cleared
    */
    GS_eventHandlingFnPtr = adjustIntProcessKey;
    /* Initialise the proposed value to the current value. */
    GS_proposedInt = *(int *)(nodePtr->valuePtr);
    /* Because we know that the first value displayed is the
       current value, tag it with " **". */
    sprintf(GS_viewString, " = %d **", GS_proposedInt);
}
```

```c
void adjustIntProcessKey(MenuNode *nodePtr, char key)
{
    /*
    If it is a return (DOWN) key, accept the change
    If the key is backspace (UP), cancel the change
    If the key is a space (SCROLL), display the next possible value.
    In cases in which we are finished processing then set the
    GS_eventHandlingFnPtr back to NULL so that the menu keys
    will be processed normally again.
    */
    switch (key)
    {
    case DOWN_KEY:
        *(int *)(nodePtr->valuePtr) = GS_proposedInt;
        break;
    case UP_KEY:
        GS_eventHandlingFnPtr = NULL;
        /*
        We have already updated the display, so we want to
        skip the following update by returning now.
        */
        return;
    case SCROLL_KEY:
        /*
        This is the one place in the function where it is necessary
        to check which particular integer we are adjusting. If
        there were a great number of integers, we could consider
        adding a field to the MenuNode to store this information
        */
        if (nodePtr == &mSetPressure)
        {
            GS_proposedInt = (GS_proposedInt + 1) %
                            (MAX_TRAINER_PRESSURE + 1);
        }
        else if (nodePtr == &mSetBrightness)
        {
            GS_proposedInt = (GS_proposedInt + 1) %
                            (MAX_TRAINER_BRIGHTNESS + 1);
        }
        else
        {
            assert(FALSE);
        }
        break;
    default:
        ; /* ignore all other keys */
    }
```

```
    sprintf(GS_viewString, " = %d", GS_proposedInt);
    /* Tag the value if it is the currently selected value. */
    if (GS_proposedInt == *(int*)(nodePtr->valuePtr))
    {
        strcat(GS_viewString, " **");
    }
}
```

The alternative to using GS_eventHandlingFnPtr is having selectInt() not return to the event loop at all but read the events itself. The following function would perform this task, and for this small example it would actually work.

```
void selectInt(MenuNode *nodePtr)
{
    char key;
    GS_proposedInt = *(int *)(nodePtr->valuePtr);
    while(1)
    {
        if (kbhit())
        {
            key = getch();
            adjustIntProcessKey(nodePtr, key);
            if (key == UP_KEY)
            {
                return;
            }
            menuUpdateDisplay();
        }
    }
}
```

The problem with this function is that if control does not revert to a central event loop while the user adjusts the current value, that task will not get an opportunity to process events other than the ones allowed for in the selectInt() function. If I added a single Exit key that would exit the menu from any point, I would have to change code in the main event loop. I would also have to update selectInt(). This would not be a satisfactory arrangement, especially because selectColor() would probably be coded the same way, leading to three updates. I am going to such pains to illustrate this mistake because it is so common.

A couple of other functions accessible from the Details menu display the size and model of the trainer. They have a simple job of displaying the menu with the extra information tagged on to the end of the menu line.

```
void displaySize(MenuNode *nodePtr)
{
    sprintf(GS_viewString, " = %d", TRAINER_SIZE);
}

/*
Currently there is only one model of this trainer. It is called Speedy, so we
can hardcode this piece of information in  the leaf for details/model.
*/
void displayModel(MenuNode *nodePtr)
{
    strcpy(GS_viewString, " = SPEEDY");
}
```

The exitProgram( ) function, accessible from the top level of the menu, provides a trivial way of escaping from the menu program.

```
void exitProgram(MenuNode *nodePtr)
{
    exit(0);
}
```

4.8.2.2 The View

Given any node, I can display the location. I name the location by naming the whole path from the top level, so the submenu Sole under Color appears as follows on the display

```
Color>Sole
```

The function menuUpdateDisplay( ) constructs this string by navigating up the tree until the top is reached, appending strings as it goes. This string is called the path because it bears a resemblance to a file system directory path. The textDisplay( ) function, which implements the physical layer, is responsible for displaying the constructed string. On the companion disk, the example implements a version of textDisplay( ) that simply displays the string on your PC's screen.

The GS_viewString is appended to the end of the path before printing. This is a useful way of inserting data associated with some of the leaves. This is view-only data, because the information is in string form and cannot be interpreted as easily as the information in the model, such as the values of settings and colors. The view data is never used for any purpose other than updating the display. I assume that the model data is used by the rest of the trainer software to control color, brightness, and air pressure.

```c
void menuUpdateDisplay(void)
{
    MenuNode *menuNodePtr = GS_currentNodePtr;
    char displayString[MENU_TEXT_LENGTH+1];
    char stringSoFar[MENU_TEXT_LENGTH+1];
    /*
    displayString would not need to be initialised here except for
    the case in which we are in the main menu and the while loop
    below will has zero iterations.
    */

    strcpy (displayString, menuNodePtr->string);
    strcpy (stringSoFar, menuNodePtr->string);
    /*
    Ensure that the last character is NULL. If this is
    overwritten, the string was too long and there is a bug
    somewhere in the program. It will be detected in the assert
    below.
    */
    displayString[MENU_TEXT_LENGTH] = '\0';
    /*
    This loop will concatenate all the names of the menu nodes,
    placing a >-in between.
    */
    while (menuNodePtr->up != NULL)
    {
        strcpy(displayString, menuNodePtr->upPtr->string);
        strcat(displayString, ">");
        strcat(displayString, stringSoFar);
        strcpy(stringSoFar, displayString);

        menuNodePtr = menuNodePtr->upPtr;
    }
    /*
    If a string was set in GS_viewString, append it to
    the end of the display string.
    */
    strcat(displayString, GS_viewString);
    /* Ensure that the string never overruns */
    assert(displayString[MENU_TEXT_LENGTH] == '\0');
    /* Print the string on the display. */
    textDisplay(displayString);
}
```

4.8.2.3 The Controller

The function menuProcessKey() is the entry point for the key events that are received. All menu manipulations start and end here. The global static GS_eventhandlingFnPtr points to a function that may process key events if the user enters particular leaves. This is a useful tool for directing events once an action function has decided that some event processing must be performed independent of the menu navigation.

```
void menuProcessKey(char key)
{
    /*
    Clear this string, any node that requires it will set it.
    */
    GS_viewString[0] = '\0';
    /*
    If a event-handling function is set up, let it process the
    key; otherwise, the key is used to navigate the menu.
    */
    if (GS_eventHandlingFnPtr)
    {
        GS_eventHandlingFnPtr(GS_currentNodePtr, key);
    }
    else
    {
        switch (key)
        {
        case DOWN_KEY:
            menuDown();
            break;
        case UP_KEY:
            menuUp();
            break;
        case SCROLL_KEY:
            menuNext();
            break;
        default:
            ; /* ignore keys that can not be processed */
        }
    }
}
```

In order to make the menu a stand-alone executable program, the module menumain.c implements a loop capturing all keyboard input and passes it to the controller (i.e., the menuProcessKey() function). The menumain.c module also defines textDisplay(), which displays the output on the PC screen.

4.8.3 Advantages of Table-Driven Code

Depending on the types of queries and commands you wish to use, you may want to add further facilities to your menu. If you add further menu functions or take away existing features, you will find that the table of nodes is the piece of code that links them. Updating this table can change the menu's behavior dramatically. Structuring the code in this way allows the programmer to see the flow of control easier than if he has to follow several levels of nested `switch` statements or `if-else-if` constructs.

If the table you construct to solve your problem grows large and unwieldy, you determine if there is a common theme being repeated. If so, it may be possible to make the table processing more powerful, allowing the table to shrink.

If the menu were implemented using the FSM approach, I would have used one line of the table for each transition between nodes. Instead, I used one line of the table for each node. The table using the FSM would have taken 18 table entries instead of 11. The memory saving is irrelevant in comparison with the difficulty of managing and maintaining large tables. If I had used one structure per transition, I would have also lost the convenience of having slots to store the name of each node and the `valuePtr` field. Having places to put things is one benefit of well-structured code and tables.

Experiment with table contents until you find a structure that matches your needs.

Chapter 5

Graphics

Embedded systems have traditionally used low-power processors and nongraphical user interfaces. This is changing for several reasons. Small, flat graphical displays are more suitable for many embedded applications than bulky cathode-ray-tube screens. The price of these flat-panel screens is dropping, even as the quality is improving because of the popularity of laptop computers. The processing power required to drive these displays is also getting cheaper. Palmtop computers are influencing the price of smaller displays, such as the quarter-VGA resolution screens.

Several design advantages are pushing developers toward graphical interfaces. Many products are marketed on the number of features they provide. If each new feature leads to a new dial or switch on the front panel, the cost of that panel rises. A graphical interface can hide controls when they are not needed, so although the number of features per square inch of front panel space is increasing, the GUI need not look more cluttered. However, careful design is required to prevent important features from being too difficult to access.

In many devices, a software upgrade can increase the ability to control a process. New front panels are often supplied with upgrades because the user needs some means to access new features. If the original system has a GUI, the new controls can be presented there, requiring no hardware upgrade. If the software upgrade is possible by serial download from a PC, the upgrade cost can approach the cost of a desktop application's software upgrade. This gives manufacturers a competitive edge if the front panel cost is high relative to the software distribution cost. Graphics screens may have a higher initial cost, but may be cheaper through the life of the system than corresponding custom control panels.

This chapter discusses how to implement such graphics systems. I will revisit the ergonomics issues of graphics displays in Chapter 9.

5.1 The Software Levels

Coding an entire GUI is intimidating. Fast graphics need special knowledge of the exact hardware configuration. Complex interactive graphics demand a set of graphical objects that can be reused in many dialogs. *Dialog* is the term I use for a display configuration or layout; the term *screen* is too ambiguous. On the desktop, a dialog would normally be a single window, but embedded systems rarely have desktop-style windows with overlapping and scrolling.

Modern desktop applications are almost never written completely by one software vendor. It is one area in which code reuse is widespread. Device drivers handle the lowest levels of putting pixels on the screen. Drawing libraries provide the functionality to draw lines, curves, bitmaps, and text. Higher level object-oriented libraries supply controls, such as buttons, menus, sliders, and tick-boxes, and support screen real-estate management with windows. The code to control these facilities is often automatically generated by a GUI builder, allowing the developer to drag and drop the graphics and controls into a window.

The higher level libraries that support objects manage the events and refresh the display. I will outline the features that can be implemented at this level. It is up to you to decide if you need any of this functionality, and then you can find out if it can be bought from a third-party vendor. If only some parts of this functionality are required, you may be able to implement it yourself.

5.2 Choosing the Set of Primitives

In this section, I describe the features that may be supplied by a library of drawing primitives. The functions should be written with direct access to the hardware because wasted CPU cycles at this level are magnified many times when you render complete screens.

Algorithms for drawing lines and arcs are covered in Foley et al. (1996). If you use a VGA-compatible display, Abrash (1996) covers all the low-level bit twiddling and optimizations you might ever want. If you must write a low-level library yourself, at least you can use the library again if you use that hardware on another product. The graphics code at this level rarely has any dependencies on the specific application.

One thing to be aware of if you are considering buying into VGA technology is that the people who sell the chips tend to sell vast amounts to very few customers. Those customers, as you might imagine, build PC compatibles. So all the programmers who use these devices can use a single BIOS call to initialize the adapter. The embedded programmer does not have this luxury and must initialize each register individually. This sounds straightforward, but few people write this type of code. The initial state of the registers for each mode is rarely documented properly. You have been warned!

5.2.1 What Do I Want to Draw?

What primitive drawing functions might you want? The typical set includes drawPixel(), drawBox(), drawLine(), drawArc(), drawText(), and drawBitmap(). There may be variations to allow for arrowed lines, numerous formats for the bitmaps, and specific fonts for rendered text. Some can be written in terms of the others. For example, the box can be drawn as a collection of horizontal lines. However, the hardware often supports faster ways of producing filled rectangles, in which case you should bypass drawLine().

Many attributes are required to draw something as simple as a box. Is it filled? What is the line thickness? What color is it? Are the corners rounded? These questions could be mapped into a large number of parameters to the drawBox() routine. To avoid long lists of parameters that would consume CPU cycles, as well as require more work on the part of the programmer, most libraries allow a pointer to a graphics context to be passed to each drawing function. The context defines many of the parameters described earlier. If many similar boxes are to be drawn, the context need not be changed between calls. If one attribute, such as color, changes, the context can be altered for that single attribute before the next call is made. A context can be shared among different drawing primitives. Some of the attributes will not always apply, such as the filled attribute when drawing a line. In those cases, the redundant attribute is ignored.

Font support is important for managing text of various sizes, although you are unlikely to need the range of typefaces used by desktop applications. Scalable fonts require more storage initially but can generate fonts of any size. In most embedded systems, all the font sizes required are known at compile time. This means that the extra work required to implement scalable fonts is rarely justifiable. The one case in which this may be warranted is if the interface allows a magnify function. If, for example, a map with labels is magnified by an arbitrary amount, the labels will also need to be scaled by that amount. Even in such a case, a set of fixed sizes may suffice if you are prepared to use the nearest fit.

Bitmaps are stored in many formats. On the desktop, they are typically stored in files. If you are avoiding a file system in your application, you may want to use a format that can be trivially compiled into your application. For example, the X-bitmap (or .xbm) format stores the image as an array declared in C syntax. Desktop applications often read this format from a file, but it can also be compiled as a module.

5.2.2 Support for Flexible Drawing

For each function listed earlier, the arguments would typically describe the location at which the image will appear. However, there may be several alternative drawing areas on which you could put the image. Some video controllers allow for several virtual screens, only one of which is visible at a time. You may also wish to draw a bitmap somewhere in memory that would be copied to the screen later. You can also implement separate coordinate systems. An origin and a scale for the *x* and *y* coordinates can be set, which allow drawing to be performed in the units relevant to the application rather than in pixels. This is particularly useful for graphs. If on a bar graph each pixel represents 100 revolutions per minute (RPM) and each motor speed is spaced by 20 pixels, the *x* scale is set to 0.01, and the *y* scale is set to 20. The origin is set to the origin of the graph on the screen. The line representing the speed in the third motor as 4,000 RPM can now be drawn with `drawLine(3, 0, 3, 4000)`, which draws a line from the *x*-axis to the height representing 4,000 RPM, which is 4,000/100 = 40 pixels. When redrawing the same diagram at different sizes at different times, the application can change the scale and then draw the diagram with the same coordinates as before.

Moving the origin is also useful if several objects are to be drawn inside one window or container. Moving the origin allows the entire group of graphics to be drawn in a new location without calculating a new location for each one. Each graphic is simply redrawn with the same arguments as before, and the new origin causes them to appear in a new location.

You may also want a clipping feature. Graphics are clipped when their appearance is limited to an area on the display, usually a rectangle, as seen in Figure 5.1. This is

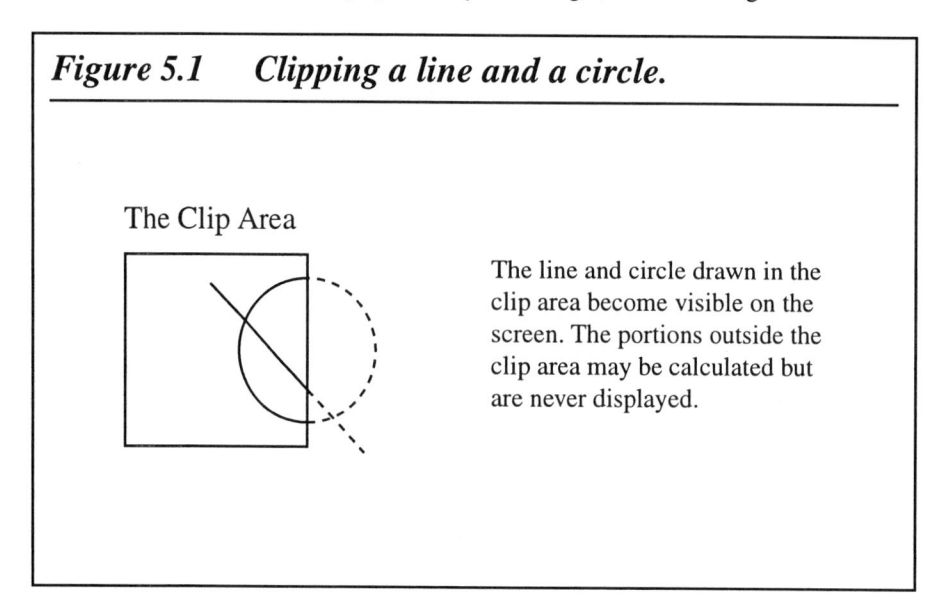

Figure 5.1   Clipping a line and a circle.

The Clip Area

The line and circle drawn in the clip area become visible on the screen. The portions outside the clip area may be calculated but are never displayed.

useful in a number of circumstances. The graphic may be inside a container or window, and you may want to limit the user's view into that container; the rest of the display may be designated for other information. At other times, the program may want to refresh one area of the screen without having an impact on any other part, since the rendering algorithm may be rendering only the objects that overlap the clip rectangle.

There are three types of clipping. At the highest level, a single shape, such as a line, can be checked to see if it is completely outside the area and removed from the list of objects to be drawn. The second level is applied when the primitive is called to render the shape. A new shape can be calculated to remove the portion outside the clipping area. For example, a shorter line than the original may be calculated, removing the portion of the line outside the clipping area. Another example would be a circle that is truncated to form an arc. This is known as preclipping. Postclipping is implemented by calculating all the pixels in the shape and then checking that the pixel is inside the clipping area just before rendering it. Postclipping is less efficient but sometimes suitable if implemented in hardware.

5.3 The Next Level Up — Do You Need Objects?

The interface to the primitives described earlier is functionally oriented. Only a minor amount of state information, such as the current drawing color, is stored in the context between calls. This is shared across all calls and is not stored as per-object state. In terms of the MVC paradigm, calls to such functions represent the view of the user interface. You could write a routine to paint a scene on the display with a series of calls to these primitives. When a different screen is needed, the display is blanked, and a different routine could contain the primitive calls to paint a new masterpiece. This is the same structure used in programs that conduct a text interaction using `printf()` calls. Any new information is simply output, and the old information is overwritten or scrolls out of the way. The situation is less simple when you change a part of the display that has already been rendered. Why not erase and redraw everything? Speed is one answer. Also, the flicker it produces could lead to early blindness or insanity. More importantly, the information required to construct the whole scene may not be available from one place. The information must be gathered from many parts of the program, leading to maintenance problems. A change in a data structure in one area leads to changes in the code needed to draw a scene in many other places.

So what is the alternative? A model can be built by designing structures that describe each box, line, button, slider, or container on the display. By maintaining these structures, previously drawn graphics can be redrawn with one or more attributes altered. You can purchase an object-oriented graphics library from a third party, or if your needs are simple, you can write one yourself. The example later in this chapter implements a simple object-oriented graphics library that provides several simple shapes, text, buttons, and containers.

Implementing a general scheme that manages a structure for each screen entity creates a certain amount of overhead. On a simple embedded system, this overhead may or may not be justified. If the display layout does not change much and there is little movement, this extra level of functionality may not be necessary. If the display is used for output only and is not interactive, you can probably get by without an object-oriented layer. However, if the user interacts with individual controls, you will want to implement an object-oriented model to control the events. If parts of the display are dedicated to separate functions that behave independently, you will want containers to define these areas and to allow them to be displayed and hidden. Hopefully, by the time you finish this section, you will understand how easy it is to apply an object-oriented structure to such graphics and the considerable advantages it offers.

Objects allow you to manipulate an image after it is drawn.

Figure 5.2 shows the levels of software that have been described in the previous sections. The top level is the application code, which varies from program to program. This level creates objects and manipulates them. When the refresh algorithm is applied, the data stored in objects is used to construct the calls to display the view of the objects. It is not necessary to refresh all the objects at the same time. The significant difference between what happens at the object level and what happens at the primitive level is that calls to the object level always record the parameters. Calls at the primitive level render the shape but do not store any data, except for calls that update the context.

The next few sections will develop objects that can refresh themselves and are managed by containers. The functionality allowing objects to overlap each other or to clip at the borders of containers is not implemented. In many applications, the programmer has enough control over the positions of objects that these features are not necessary. Building a more powerful set of objects leads to a more processor-hungry implementation, making it less applicable to small embedded systems. All the code in these sections comes from the `multi` (short for multidialog) program on the companion disk.

5.3.1 Structures to Define Graphic Objects

I want to represent lines, boxes, text strings, and circles. Each one requires its own structure to store data unique to that graphic. The box may be filled or not. The text object must store the characters that it is to display. The line object must store start and end points.

Several attributes, such as location and color, are common. If I extract them into another structure, I can include this new structure in each of the mentioned structures. By creating this `Drawable` structure, you can write functions that use the area of the graphic, without worrying about which type of graphic is manipulated. Another useful abstraction is a single structure for the `Area`.

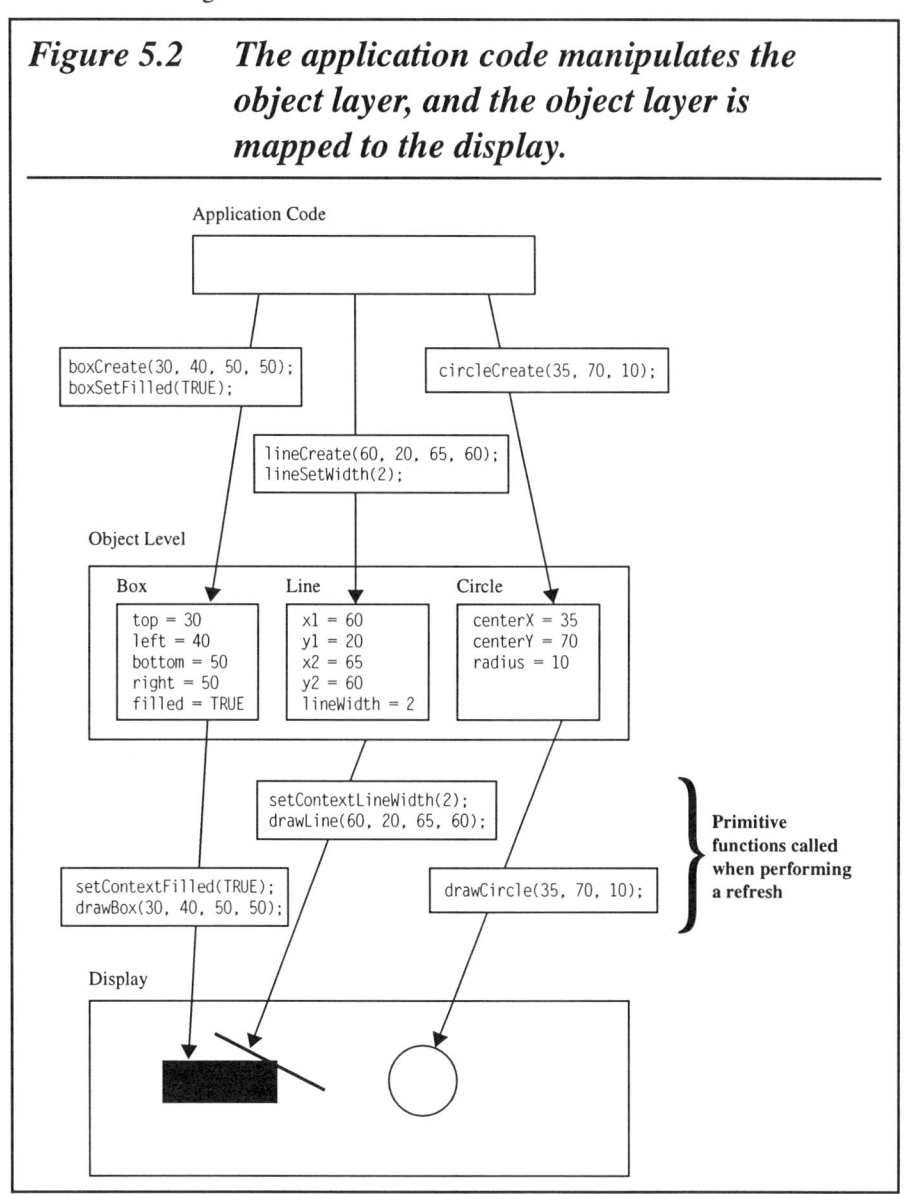

Figure 5.2 ***The application code manipulates the object layer, and the object layer is mapped to the display.***

```
Application Code
```

```
boxCreate(30, 40, 50, 50);
boxSetFilled(TRUE);
```

```
circleCreate(35, 70, 10);
```

```
lineCreate(60, 20, 65, 60);
lineSetWidth(2);
```

Object Level

Box	Line	Circle
top = 30 left = 40 bottom = 50 right = 50 filled = TRUE	x1 = 60 y1 = 20 x2 = 65 y2 = 60 lineWidth = 2	centerX = 35 centerY = 70 radius = 10

```
setContextLineWidth(2);
drawLine(60, 20, 65, 60);
```

```
setContextFilled(TRUE);
drawBox(30, 40, 50, 50);
```

```
drawCircle(35, 70, 10);
```

} Primitive functions called when performing a refresh

Display

For the `Area`, I store the top-left point and the bottom-right point. Although storing the width and height may seem more intuitive, several calculations, such as checking for overlap, are simpler using the bottom-right point representation.

The following structures define the `Area`, the `Drawable`, and some of the shapes that I require.

```
typedef enum {CONTAINER, BOX, CIRCLE, LINE, TEXT} DrawableType;
typedef struct
{
    int left;
    int top;
    int right;
    int bottom;
} Area;
struct drawableStruct
{
    Area area;
    int color;
    DrawableType type;
};
struct boxStruct
{
    Drawable drawable;
    Boolean filled;
    int fillColor;
};
struct circleStruct
{
    Drawable drawable;
};
struct lineStruct
{
    Drawable drawable;
    int x1;
    int y1;
    int x2;
    int y2;
};
struct textStruct
{
    Drawable drawable;
    char *string;
};
```

The structure names are `typedef`ed to be type names in `shapes.h` according to the following definitions. This avoids having to use the keyword `struct` each time one of these structures is referenced.

```
typedef struct drawableStruct Drawable;
typedef struct boxStruct Box;
typedef struct circleStruct Circle;
typedef struct lineStruct Line;
typedef struct textStruct Text;
```

The circle structure is empty because the radius and center can be derived from the `Area` structure stored in the `Drawable`. On the other hand, the line structure contains some redundant information, because the area itself cannot unambiguously identify the line. Once the rectangle containing the line is defined, it is still necessary to identify that the line is from the top-left to the bottom-right or from the top-right to the bottom-left. I dislike redundant information because the bugs caused when the two forms become inconsistent can be difficult to track down. But in this case, it is unavoidable.

The `Drawable` structure is included as the first field of each graphic structure. This allows you to access the `Drawable` using a pointer to one of the other graphics by simply casting it to a pointer to a `Drawable`. I hide this cast inside a macro, which can then apply to any of the shapes defined earlier.

```
#define GET_DRAWABLE(d) ((Drawable *)(d))
```

This prevents the user from being aware of the cast. It also allows you to write a debugging version of the macro if you want to sanity-check the contents of the drawable each time you cast to it.

```
#define GET_DRAWABLE(d) (checkDrawable((Drawable *)(d)))
```

`checkDrawable()` checks each field of the drawable structure to see if it is within the bounds for that field and exits with an error code if a field is out of range. This does not guarantee that you will catch an illegal cast, but it is better than doing nothing.

By extracting common data into a single data structure, you can implement a simple form of polymorphism.

The `typedef`s for the structures are separate from the structure definitions. Thus, you can make the types visible in a header file while keeping the structures in a `.c` file. This also implements *opaque types*, which allows the caller to hold pointers to the structures without having access to the members of the structures themselves. You can ensure that any changes made to data stored in the structures is via the functions provided.

5.3.2 Memory Management and Initialization

Declaring these structures statically or on the stack is not suitable for graphics applications. If the structures are declared on the stack, they cease to exist when the function exits. This means that the object will have a short life. For a desktop application, many programmers simply allocate these objects on the heap. In embedded systems, which may run continuously for many months, the heap can be the source of problems. In C, malloc() and free() allow blocks of bytes to be allocated from the heap and returned to it. If these functions are called often, and for blocks of varying size, heap fragmentation eventually renders the heap unusable for large allocations, and the program fails. A heap is fragmented when the chunks of memory allocated are scattered throughout the heap's memory space. The remaining space is broken into so many small pieces that allocating a large block is impossible, even though a large percentage of the memory is unused (Figure 5.3). Fragmentation should not be confused with a memory leak. A leak occurs when memory is allocated but never freed. If the offending function is run only occasionally, an allocation failure may not occur until the system has been operating for some time. Because the allocation that cannot be satisfied may be unrelated to the allocation that caused the problem, finding leaks can be quite difficult. The root cause of a leak is generally the piece of code that performs an allocation. Leaks can usually be remedied by freeing memory at the appropriate time. Fragmentation is more difficult to avoid because it is an intrinsic property of a heap that allocates blocks of various sizes that cannot be moved once allocated.

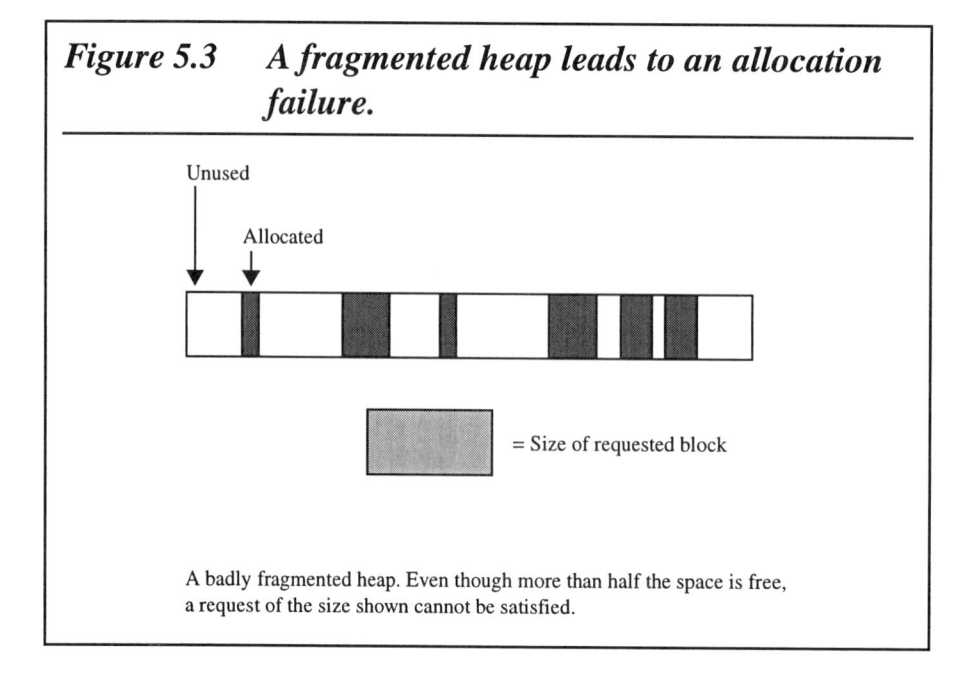

Figure 5.3   *A fragmented heap leads to an allocation failure.*

A badly fragmented heap. Even though more than half the space is free, a request of the size shown cannot be satisfied.

For these reasons, many embedded programmers eschew even the most cautious use of malloc() and free(). This is not an unreasonable approach. You simply decide on all the structures and buffers that your program may need and provide for them up-front. They can be declared statically, and the compiler will set aside space for them.

You should understand that by not using the heap, your memory requirements are greater than an equivalent program using the heap in the absence of a leak. Consider a program that needs 10 settings structures to reflect settings that the user can change, as well as related information, such as limits, resolution of change, and data specific to the type of input device used to change the setting. If the largest number of settings in use at a time is three, the memory consumption is three times the settings structure size. If all 10 are allocated statically, the memory consumption is 10 times the size of the setting structure. So by allocating all structures statically, you settle for the worst-case memory consumption, but you have a guarantee of no leaks. In embedded systems, this may be acceptable because the number of elements on a display is often limited by the physical control panel. However, this changes dramatically when you start to use graphics.

If you do not want to use malloc() and free(), then declaring the structures statically is an option, but it has its drawbacks. You encounter two problems with declaring the structures statically. First, each structure must be given a unique name in the global scope. If there are many objects, it may be difficult to find meaningful names for them. Second, the structures exist in an uninitialized state until the program has enough information to set initial values. There is a danger that the program may use a structure before initialization, with unpredictable results.

You can get some of the convenience of heap allocations and none of the dangers if you use the same approach taken in the shapes.c module. A piece of memory is set aside by allocating a static array of unsigned chars. The salloc() routine allows memory to be allocated from this block, but it is never freed. If the block is used up, an assert prevents the program from running. The salloc() routine is used only during start-up so that any problems may be found as soon as the system is run. This leads to problems being discovered in test, not in the field after release. You can enforce this by adding a function that disables salloc() after start-up is complete. The salloc() routine is local to the shapes.c module, preventing any other part of the system from using it. You may choose to make your pool more globally accessible. But for this application, I chose to allow only shapes on my heap, nothing else. I could not decide whether salloc should mean static allocate or shape allocate, so you can pick either. salloc() can be made more sophisticated by allocating blocks the size of the largest shape. Then, if blocks are freed and reallocated, fragmentation can be avoided. Some more possibilities in this area are explored in Chapter 7.

```
#define SALLOC_BUFFER_SIZE 5000
unsigned char GS_sallocBuffer[SALLOC_BUFFER_SIZE];
int GS_sallocFree = 0;
void *salloc(int size)
{
    void *nextBlock;
    if(GS_sallocFree + size > SALLOC_BUFFER_SIZE)
    {
        return NULL;
    }
    nextBlock = &GS_sallocBuffer[GS_sallocFree];
    GS_sallocFree += size;
    return nextBlock;
}
```

Now that you can allocate the memory for the shapes, you will want to be able to initialize them at creation time to avoid the possibility of using an uninitialized object. This is performed by a number of functions to create and initialize an instance of each structure. Because each structure contains a Drawable, you must call drawableInit() on that part of the structure. I will show the creation functions for Box and Text, the others follow much the same form.

```
void drawableInit(Drawable *drawablePtr, DrawableType type,
                  int left, int top, int right, int bottom)
{
    drawablePtr->color = BLACK;
    drawablePtr->area.left = left;
    drawablePtr->area.top = top;
    drawablePtr->area.right = right;
    drawablePtr->area.bottom = bottom;
    drawablePtr->type = type;
    drawablePtr->parentPtr = NULL;
    drawablePtr->nextContainedPtr = NULL;
/*
The object is not initially dirty. It will be marked dirty when
it is added to a container
*/
    drawablePtr->nextDirtyDrawablePtr = NULL;
    drawablePtr->dirty = FALSE;
    drawablePtr->oldParentPtr = NULL;
}
```

```
Box * boxCreate(int left, int top, int right, int bottom)
{
    Box *boxPtr;
    boxPtr = salloc(sizeof(Box));
    assert (boxPtr != NULL);
    drawableInit(&boxPtr->drawable, BOX, left, top, right, bottom);
    boxPtr->filled = FALSE;
    return boxPtr;
}
Text * textCreate(int left, int top, char *string)
{
    Text *textPtr;
    int right;
    int bottom;
    textPtr = salloc(sizeof(Line));
    assert(textPtr != NULL);

#ifdef USING_BGI
    right = left + textwidth(string);
    bottom = top + textheight(string);
#endif
    drawableInit(&textPtr->drawable,TEXT,left,top,right,bottom);
    textPtr->string = string;
    return textPtr;
}
```

5.3.3 *Container Hierarchies*

It is useful to put the shapes into containers that can then carry them around. Compound objects can then be moved as a single unit or inserted and deleted by a single function call. Containers also provide their own coordinate space. The location of each object within a container is relative to the container's origin, not relative to the display origin. To draw a container, you must first draw the background color then draw each element within the container. Because containers can hold other containers, this drawing algorithm can become recursive.

By breaking the screen into several regions, each of which is occupied by a container, refreshing the display can become more efficient. If the container in one region is replaced by a new container, only that region will be redrawn. Containers in other areas remain unchanged.

One root container occupies the whole display. The root's subcontainers and all descendants are displayed while objects not connected to the root are invisible. Such invisible containers can be useful places to build up dialogs, which will possibly be attached to the root later. A visible flag in the container records whether it is attached to the root. This avoids following parent pointers to the top of the tree each time a decision is made to draw an object.

> **Containers are a simple form of a window with no frame. They allow us to control areas of the screen independently.**

In the code supplied, the coordinates of any object are the coordinates within the parent container. When the object is being drawn, you must calculate the absolute coordinates in order to render the object on the display. For efficiency, the containers maintain their absolute position, as well as their position within their parent (remember, containers are contained in other containers). The absolute position is meaningless unless the object is connected to the root container and has a position on the display.

The `Container` structure is:

```
struct containerStruct
{
    Drawable drawable;
    Drawable *containedListPtr;
/*
The absolute location is maintained to optimize drawing the
contained objects.
*/
    int absoluteLeft;
    int absoluteTop;
    Boolean visible;
};
typedef struct containerStruct Container;
```

The container uses `Drawable` to control its area, as do the other shapes. The `containedListPtr` points to the `Drawable` part of the first child of this container. Each child is then linked to the next with a null pointer as a terminator.

The containers implemented in the example code do not clip the graphics, so the contents may extend outside the boundaries of the parent container. It is up to the application to ensure that this does not happen.

Because a container has a color stored in the drawable structure, each container can have a different background color. This makes the container's boundary obvious to the user. If this is not the desired effect, make the colors the same as the root container. There is a hierarchy of containers because containers can hold other containers. The hierarchy changes dynamically at run time as the shapes are added and removed from containers. Figure 5.4 shows the container hierarchy when the graphs screen is visible. You may wish to run the `multi` example program on the companion disk so that the objects identified on this tree can be seen on your PC. Objects lower in the hierarchy are physically inside their parents as you look at them on the display. Note that the names do not directly correspond to type or variable names in the code. That is because some objects have temporary names, such as `buttonPtr`, that would be meaningless in this context. I also omitted some text objects to keep the diagram simple.

With the creation functions described in the previous section, shapes can be created and added to containers without having to set aside a unique name for them. This is useful for some objects that, once added to the `Container`, do not require further manipulation. For example, once the following function returns, no unique name is reserved for the `Text` structure, although `Text` continues to exist and will be visible whenever the parent `Container` is visible.

```
void addText(Container *cPtr)
{
    Text *t1 = textCreate(20, 40, "Hello World!");
    containerAddTo(cPtr, t1);
}
```

5.4 Refreshing the Display

Now that you have objects you can draw and containers to hold them in, how do you decide when to refresh them? If you tell the outermost container to redraw all of its children, the entire screen is redrawn. This works but is extremely inefficient if such a redraw is performed for every change. I need a way to identify the areas that need to be refreshed.

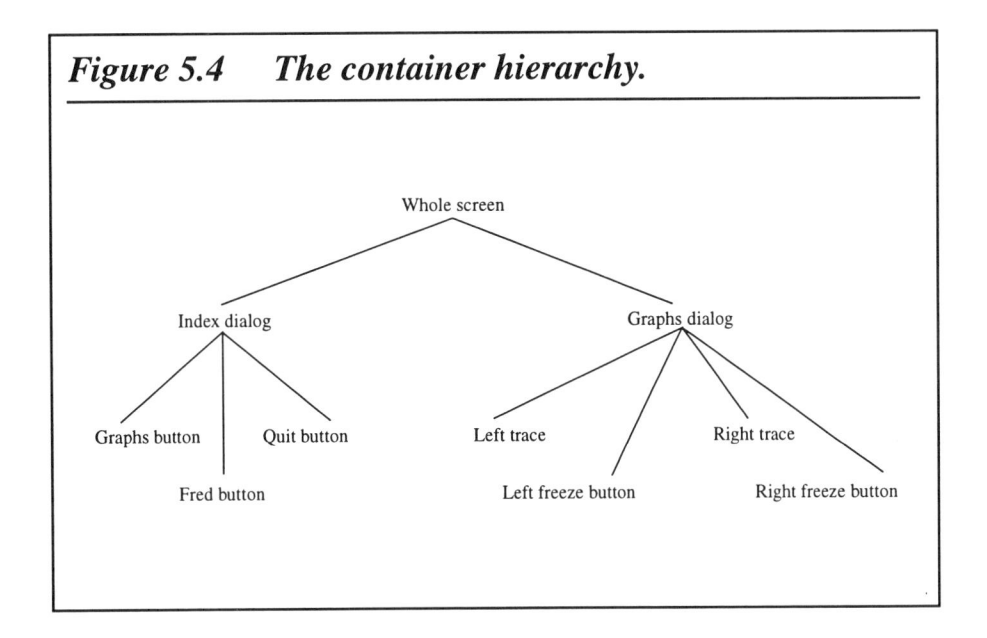

Figure 5.4   **The container hierarchy.**

Figure 5.5 shows an object being moved. At the application level, the request is simply to move the man to a new position. To refresh the display, two actions must occur. First, the man image must be removed from its current position. Second, the image must be drawn in its new position. This seems obvious, but note that, at the primitive level, the man cannot be moved. He must be completely redrawn. Other images on the display must not be corrupted by this change. Ideally, the only areas of the display refreshed are the old position of the man and the new position of the man. These are known as dirty areas.

Refreshing the dirty areas is not as trivial as you might think. Consider erasing the man shown in Figure 5.5. He must be redrawn in the background color of the parent container. If you want to allow overlapping objects, you must reproduce objects that are uncovered. I will describe two approaches: refresh by dirty objects and refresh by dirty area. The first is simple and fast; the second is more complex but allows far more flexibility.

5.4.1 Refreshing by Object

Refreshing by object involves tracking the objects that have been altered by calls to the functions that update their structure. It does not track second-order modifications, such as an object that is uncovered by removing another object that was hiding it. Because overlapping objects are not supported, this technique is not appropriate for drag-and-drop environments or for animation. However, refreshing by object is more than sufficient for simple GUIs in which the positions of the objects can be predicted in advance and the amount of movement is restricted.

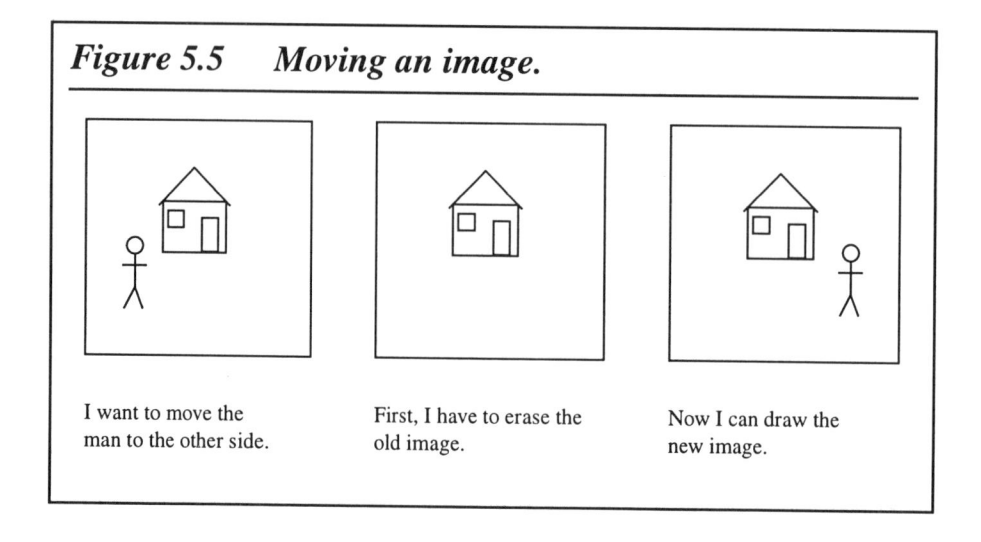

Figure 5.5   Moving an image.

I want to move the man to the other side.

First, I have to erase the old image.

Now I can draw the new image.

The changes are made with modifying functions, such as `boxSetPosition()` or `drawableSetColor()`. You associate a dirty flag with each object. When the object is modified, the flag is set to dirty, indicating that this object's area needs to be cleaned. Unfortunately, objects not only must know how to draw themselves, they must also know how to erase themselves. An object can erase itself by drawing over the old image in the background color of the container. Each object must then have a pointer to its parent to allow it to access the background color. But what if the parent moves? If you try to erase an object whose parent has moved, you will erase in the new parent position. This is a wasted operation because the parent is being completely erased and redrawn. As an optimization, the shape checks to see if the parent is dirty. If it is, it does not redraw itself.

The next possibility is that the object is removed from one parent and placed in another. The old parent must be recorded to allow the old image to be removed. If the old parent is NULL, it is assumed that it did not have a parent and there is no image to erase. For erasing, you must check if the old parent is dirty. If so, do not erase the object, because the old parent will look after removing the old image.

Erasing the old image could be performed by simply drawing the entire area in the background color, but this has drawbacks. For example, an object, such as a line, may not overlap with any other shapes, but the rectangle that surrounds it covers a larger area and may overlap with other objects. Another problem is with arrangements, such as a box that surrounds a piece of text. If the box changes color, you do not want to erase the text also.

You must implement an erase function that is similar to the draw function. But it redraws the object in the background color to make it invisible rather than in the foreground color to make it visible. Enough information should be stored each time the object is drawn so that it can be erased later. One pointer per object is needed to allow you to chain a list of dirty objects together. This avoids having to search the entire container hierarchy looking for dirty objects when the refresh is performed. So the structures grow as follows.

```
struct drawableStruct
{
    Area area;
    int color;
    DrawableType type;
    Container *parentPtr;
    Drawable *nextContainedPtr;
    Boolean dirty;
    Drawable *nextDirtyDrawablePtr;
    Container *oldParentPtr;
    Area oldArea;
};
```

```
struct boxStruct
{
    Drawable drawable;
    Boolean filled;
    int fillColor;
    Boolean oldFilled;
};
struct circleStruct
{
    Drawable drawable;
};
struct lineStruct
{
    Drawable drawable;
    int x1;
    int y1;
    int x2;
    int y2;
    int oldX1;
    int oldY1;
    int oldX2;
    int oldY2;
};
struct textStruct
{
    Drawable drawable;
    char *string;
};
struct containerStruct
{
    Drawable drawable;
    Drawable *containedListPtr;
/*
The absolute location is maintained to optimize drawing the
contained objects.
*/
    int absoluteLeft;
    int absoluteTop;
    Boolean visible;
};
```

Whenever an object is modified, the object is marked dirty by calling drawableDirty(). In the case of boxSetPosition(), the width and height must be conserved by changing the bottom-right point as well as the top-left point. This is an example of the many functions that make modifications to one of the shape's data members.

```
void boxSetPosition(Box * boxPtr, int left, int top)
{
    int deltaX = left - boxPtr->drawable.area.left;
    int deltaY = top - boxPtr->drawable.area.top;
        boxPtr->drawable.area.right += deltaX;
    boxPtr->drawable.area.bottom += deltaY;
    boxPtr->drawable.area.left = left;
    boxPtr->drawable.area.top = top;
    drawableDirty(&boxPtr->drawable);
}
```

The drawableDirty() function has two jobs: to set the dirty flag to TRUE and to add the object to the list of dirty objects.

```
void drawableDirty(Drawable *drawablePtr)
{
    if (drawablePtr->dirty == TRUE)
    {
        /*
        If we are already dirty then no need to proceed
        */
        return;
    }
    drawablePtr->dirty = TRUE;
    /* Append drawablePtr to the list */
    drawablePtr->nextDirtyDrawablePtr = GS_dirtyList;
    GS_dirtyList = drawablePtr;
}
```

Listing 5.1 shows the sequence of events when refreshing the display. When it has been decided to perform a refresh, displayRefresh() is called. It first erases the old images, then draws the new images, and finally clears the dirty flags. The drawableDraw() and drawableErase() functions decide which type of object it is and then call the appropriate functions. I will illustrate the draw and erase functions for Box and Container. The other shapes follow the form of the functions for Box. I used the Borland Graphics Interface (BGI) library, which is shipped with Borland C++. The primitives it supplies are self-explanatory and are typical of what may be provided with any third-party drawing library. I have wrapped any BGI-specific code in a conditional compile of USING_BGI. It is these conditionally compiled sections that would need to be rewritten to port the code to another drawing library.

All the information required to erase an object is recorded when the object is drawn. This means that when it is erased, the information relates to the last place it was drawn, not to some intermediate state. Consider the following fragment of code.

```
boxSetPosition(boxPtr, x1, y1);
refreshDisplay();
boxSetPosition(boxPtr, x2, y2);
boxSetPosition(boxPtr, x3, y3);
refreshDisplay();
```

Listing 5.1 **The** `displayRefresh()`, `drawableDraw()`, `drawableErase()`, `boxDraw()`, `boxErase()`, `containerDraw()`, **and** `containerErase()` ***functions.***

```
void displayRefresh(void)
{
    Drawable *drawablePtr;   /* for iterating through lists */
    int offsetX;
    int offsetY;
/*
First erase the old images of the dirty objects.
*/
    for (drawablePtr = GS_dirtyList;
         drawablePtr != NULL;
         drawablePtr = drawablePtr->nextDirtyDrawablePtr)
    {
        /*
        We erase the image only if there is an old parent and it is not dirty.
        Strictly speaking, we should check the parent's parent and so on, but the
        extra work would probably not pay off in terms of avoiding extra redraws.
        */
        if ( (drawablePtr->oldParentPtr != NULL) &&
             (drawablePtr->oldParentPtr->visible) &&
             (!drawablePtr->oldParentPtr->drawable.dirty) )
        {
            offsetX = drawablePtr->oldParentPtr->absoluteLeft;
            offsetY = drawablePtr->oldParentPtr->absoluteTop;
            drawableErase(drawablePtr, offsetX, offsetY);
        }
    } /* end for */
/*
The next step is to draw all of the dirty objects.
*/
    for (drawablePtr = GS_dirtyList;
         drawablePtr != NULL;
         drawablePtr = drawablePtr->nextDirtyDrawablePtr)
    {
        /*
        We draw the image only if there is a parent and it is not dirty.
        Strictly speaking, we should check the parent's parent and so on,
        but the extra work would probably not pay off.
        */
        if ( (drawablePtr->parentPtr != NULL) &&
             (drawablePtr->parentPtr->visible) &&
             (!drawablePtr->parentPtr->drawable.dirty) )
        {
            offsetX = drawablePtr->parentPtr->absoluteLeft;
            offsetY = drawablePtr->parentPtr->absoluteTop;
            drawableDraw(drawablePtr, offsetX, offsetY);
        } /* end if */
    } /* end for */
```

Listing 5.1 (continued)

```c
/*
Now that the objects have been updated, the dirty flag can be cleared.
These flags should not be cleared in one of the earlier loops because
they are used for optimizations in the drawableDraw function. Note that
the links of the chain are not set to NULL. This does not matter because
they will get overwritten if it is put on the chain again.
*/
    drawablePtr = GS_dirtyList;
    for (drawablePtr = GS_dirtyList;
         drawablePtr != NULL;
         drawablePtr = drawablePtr->nextDirtyDrawablePtr)
    {
        drawablePtr->dirty = FALSE;
    }
    GS_dirtyList = NULL;
}
void drawableDraw(Drawable *drawablePtr, int offsetX, int offsetY)
{
    /* draw should never be called for an orphan */
    assert(drawablePtr->parentPtr != NULL);
    switch (drawablePtr->type)
    {
    case CONTAINER:
        containerDraw((Container *)drawablePtr, offsetX, offsetY);
        break;
    case BOX:
        boxDraw((Box *)drawablePtr, offsetX, offsetY);
        break;
    case CIRCLE:
        circleDraw((Circle *)drawablePtr, offsetX, offsetY);
        break;
    case LINE:
        lineDraw((Line *)drawablePtr, offsetX, offsetY);
        break;
    case TEXT:
        textDraw((Text *)drawablePtr, offsetX, offsetY);
        break;    default:
        assert(FALSE);
    }
    /*
    Update the oldParent for the time when you want to erase this drawable.
    */
    drawablePtr->oldParentPtr = drawablePtr->parentPtr;
}
/*
This function decides what kind of shape we have and chooses an
erase routine based on the type field of the Drawable structure.
This function follows the same form as the drawableDraw() function.
*/
void drawableErase(Drawable *drawablePtr, int offsetX, int offsetY)
{
    if (drawablePtr->oldParentPtr == NULL)
    {
        /*
         If there is no old parent, there is nothing to erase from.
        */
        return;
    }
```

The first call to refreshDisplay() draws the box at the position x1, y1. At this time, the area and filled state of the box is recorded. Now the box is moved to x2, y2 and the box is marked dirty. The move to x3, y3 does not mark the box dirty because

Listing 5.1 *(continued)*

```
    switch (drawablePtr->type)
    {
    case CONTAINER:
        containerErase((Container *)drawablePtr, offsetX, offsetY);
        break;
    case BOX:
        boxErase((Box *)drawablePtr, offsetX, offsetY);
        break;
    case CIRCLE:
        circleErase((Circle *)drawablePtr, offsetX, offsetY);
        break;
    case LINE:
        lineErase((Line *)drawablePtr, offsetX, offsetY);
        break;
    case TEXT:
        textErase((Text *)drawablePtr, offsetX, offsetY);
        break;
    default:
        assert(FALSE);
    }
}

void boxDraw(Box * boxPtr, int offsetX, int offsetY)
{
    int top;
    int left;
    int right;
    int bottom;
    assert(boxPtr->drawable.parentPtr != NULL);
    left = offsetX + boxPtr->drawable.area.left;
    top = offsetY + boxPtr->drawable.area.top;
    right = offsetX + boxPtr->drawable.area.right;
    bottom = offsetY + boxPtr->drawable.area.bottom;
#ifdef USING_BGI
    setcolor(boxPtr->drawable.color);
    if (boxPtr->filled)
    {
        setfillstyle(SOLID_FILL, boxPtr->fillColor);
        fillRectangle(left, top, right, bottom);
    }
    else
    {
        setfillstyle(EMPTY_FILL, boxPtr->fillColor);
        rectangle(left, top, right, bottom);
    }
#endif
    /*
    We have to record all the old stuff now, so it can be used on the next erase.
    */
    boxPtr->drawable.oldArea = boxPtr->drawable.area;
    boxPtr->oldFilled = boxPtr->filled;
}
```

it already is dirty. The second call to `displayRefresh()` erases the box at x1, y1 and draws the box at x3, y3. No activity takes place at x2, y2 because the box was never rendered in that position.

At each call to `refreshDisplay()`, the dirty objects are erased from the old positions and drawn in their new positions. The list is then cleared, so any further modifications cause objects to be added to a fresh list. One optimization applied here is to skip objects on the list whose parent is dirty, because when the parent gets drawn, all the children are drawn. There is no point in drawing an object twice.

You will find it straightforward to add other objects to this scheme. An entry in the `switch` statement for `drawableDraw()` and `drawableErase()` has to be made, and the `DrawableType` enumeration has to be extended to include a new name. The new structure should include `Drawable` as its first member. A create, draw, and erase function has to be written. The functions to change the appearance of the new type of object could be anything you consider suitable, but they must call `drawableDirty()` after changing the data members.

Listing 5.1 (continued)

```c
/*
This function draws over a box in the background color of the parent and,
therefore, erases the box. The background color of the old parent is used
because the current parent may be different and the image that we are trying
to remove is in the old parent.
  If the old parent has moved or changed color, this routine would not work.
This is not a problem, because if the old parent had moved, it would be marked
dirty and this routine would never be called. The redraw of the old parent
would clean any image of the box that remained.
*/
void boxErase(Box *boxPtr, int offsetX, int offsetY)
{
   int backgroundColor;
   int absoluteLeft = offsetX + boxPtr->drawable.oldArea.left;
   int absoluteTop = offsetY + boxPtr->drawable.oldArea.top;
   int absoluteRight = offsetX + boxPtr->drawable.oldArea.right;
   int absoluteBottom = offsetY + boxPtr->drawable.oldArea.bottom;
   backgroundColor = boxPtr->drawable.oldParentPtr->drawable.color;
#ifdef USING_BGI
    setcolor(backgroundColor);
    if (boxPtr->oldFilled)
    {
        setfillstyle(SOLID_FILL, backgroundColor);
        fillRectangle(absoluteLeft, absoluteTop, absoluteRight, absoluteBottom);
    }
    else
    {
        setfillstyle(EMPTY_FILL, backgroundColor);
        rectangle(absoluteLeft, absoluteTop, absoluteRight, absoluteBottom);
    }
#endif
}
```

Listing 5.1 (continued)

```
/*
To draw a container, we simply draw the background and then draw
each of the contained drawables.
*/
void containerDraw(Container *containerPtr,int offsetX,int offsetY)
{
    Drawable *childPtr;
    int left, top, right, bottom;
/*
We draw the container's background in the coordinate space of the parent.
If there is no parent then we use 0,0.
*/
    left = offsetX + containerPtr->drawable.area.left;
    top = offsetY + containerPtr->drawable.area.top;
    right = offsetX + containerPtr->drawable.area.right;
    bottom = offsetY + containerPtr->drawable.area.bottom;
#ifdef USING_BGI
    setfillstyle(SOLID_FILL, containerPtr->drawable.color);
    setcolor(containerPtr->drawable.color);
    fillRectangle(left, top, right, bottom);
#endif
    for (childPtr = containerPtr->containedListPtr;
            childPtr != NULL;
            childPtr = childPtr->nextContainedPtr)
    {
        drawableDraw(childPtr, left, top);
    }
    containerPtr->drawable.oldArea = containerPtr->drawable.area;
}

/*
This function erases a container by drawing a rectangle in the old parent's
background color. This covers any shapes that container may contain.
*/
void containerErase(Container *containerPtr,int offsetX,int offsetY)
{
    int top;
    int left;
    int right;
    int bottom;
    int backgroundColor;
    left = offsetX + containerPtr->drawable.oldArea.left;
    top = offsetY + containerPtr->drawable.oldArea.top;
    right = offsetX + containerPtr->drawable.oldArea.right;
    bottom = offsetY + containerPtr->drawable.oldArea.bottom;
    backgroundColor = containerPtr->drawable.oldParentPtr->drawable.color;
#ifdef USING_BGI
    setfillstyle(SOLID_FILL, backgroundColor);
    setcolor(backgroundColor);
    fillRectangle(left, top, right, bottom);
#endif
}
```

5.4.2 Refreshing by Area

A more sophisticated technique is refreshing by area. I will not supply code to implement it. It is the way in which most commercially available libraries handle screen refreshes. It can support overlapping objects and maintain a two-and-a-half dimensions environment in which each object is on a unique plane, allowing it to pass in front of or behind all the other objects.

In the refresh-by-object scheme described earlier, if two objects occupy the same space, the most recently modified one appears toward the front. This is sometimes the preferred behavior, as when an object is dragged from one location to another. However, the objects that were hidden do not reappear when the object moves away. If this is a rare occurrence, an explicit call from the application to redraw a particular container may keep all elements visible. However, this is not a good general-purpose solution.

A better scheme is to record the areas at which changes took place rather than record pointers to the objects. At each refresh, I traverse this list of areas, looking for every object that overlaps with the area and redrawing it. The objects must be searched for in a back-to-front order so that objects closest to the viewer are drawn last, obscuring all or parts of the objects further back.

The efficiency of this scheme suffers if many areas are compared with many objects on each refresh. There are two important areas to optimize. If areas can be combined or removed, the number of searches can be reduced. For example, if an area is completely within another area on the list, it can be removed because it is redrawn anyway. If two areas have a large overlap, it may be worth replacing them with a single rectangle that includes both areas. The second area to optimize is the search for objects. If areas of the screen are controlled by separate containers, the number of objects that require checking may be reduced. Once it is established that a particular area does not overlap with a particular container, no search is performed on the objects within that container. This optimization depends on cooperation with the application layer, because the application layer places the objects in their parent containers. An application that chooses to place all objects directly into the root container does not benefit from this optimization. You will see later that grouping objects into containers has similar advantages when you need to locate a user event on the screen.

In this scheme, it is not necessary that objects know how to erase themselves. Because an area is being redrawn, it is first filled with the background color, then each overlapping object is drawn. If no objects occupy the area, this has the effect of clearing the area.

5.4.3 When to Refresh?

For either of the schemes I've mentioned, you must choose an appropriate time to perform the refresh. If an event takes a small amount of time, by the user's time scale, then perform the refresh at the end of an event. If an event takes longer, perhaps many seconds, you may want to provide some feedback at the start of the event or perform continuous feedback during the event. In these cases, you must call the refresh function whenever you want the display to be brought up to date.

5.5 Compound Objects

You have seen how a simple object, such as a box, can be implemented. Often, you will want more sophisticated objects, such as buttons or sliders. There are two approaches to building these objects. One is to create a container at run time and fill it with the shapes and text required to give it the desired appearance. Some of the contents then change according to user input or desired output. This approach has the advantage of not requiring any low-level graphics calls. The high-level objects already available can render themselves, and you make good use of functionality already available. The container can be moved and inserted into different containers, allowing you to treat the group of objects as a single entity.

If you have many instances of an object and require an efficient implementation, you may wish to create a first-order object for it. This means implementing a draw function and an erase function and writing a small number of modifier functions to allow the application layer to manipulate it. You could also add event handling at this level.

5.5.1 Button

The button implemented in Chapter 4 is a perfect candidate for integration here. I already have the `buttonDraw()` function, but now I also need a `buttonErase()` function. The erase function is trivial to write because there is a simple rectangle to blank, as there was with the Box. I also require a `buttonCreate()` function because I will make Button an opaque type, like the other shapes.

The enumeration of objects must be expanded to include BUTTON.

```
typedef enum {CONTAINER, BOX, CIRCLE, LINE, TEXT, BUTTON}
    DrawableType;
```

The structure describing the button requires new fields to allow it to work with the other shapes and containers. It also contains two `ActionFns`, which hold pointers to functions that are the callbacks associated with this button.

```
struct buttonStruct
{
    Drawable drawable;
    char *text;
    ButtonState state;
    ActionFn actionFnDOWN;
    ActionFn actionFnUP;
};
```

Several functions (Listing 5.2) are required to allow the application to create and modify buttons. The buttonDraw() routine breaks down into a routine that draws the button in each of its four states. A little addition and subtraction is required to calculate the polygons displayed as borders. The buttonErase() function only has to blank the rectangle.

Listing 5.2 The functions required to allow a graphics application to create and modify buttons.

```
Button *buttonCreate(int left, int top, int right, int bottom,
            char *string, ActionFn downFn, ActionFn upFn)
{
    Button* buttonPtr;
    buttonPtr = salloc(sizeof(Button));
    assert (buttonPtr != NULL);
    drawableInit(&buttonPtr->drawable,BUTTON,left,top,right,
    bottom);
    buttonPtr->text = string;
    buttonPtr->state = UP;
    buttonPtr->actionFnUP = upFn;
    buttonPtr->actionFnDOWN = downFn;
    return buttonPtr;
}
/*
Set the string that will be visible inside the button.
Note that no copy of the string is taken, so it is expected
that it will be assigned to constant strings as in:
buttonSetText(btnPtr, "hello");
*/
void buttonSetText(Button *buttonPtr, char *string)
{
    buttonPtr->text = string;
    drawableDirty(GET_DRAWABLE(buttonPtr));
}
/*
Set the button state to its initial state.
*/
void buttonReset(Button *buttonPtr)
{
    buttonPtr->state = UP;
    drawableDirty(GET_DRAWABLE(buttonPtr));
}
```

The `drawableDraw()` and `drawableErase()` functions must also be modified for the new addition. The `switch` statement that checks the type of the `Drawable` must have a case added for the `BUTTON` type.

The event management still uses the finite-state machine described in Chapter 4. The `buttonProcessFSM()` function operates almost exactly as before. The following section on locating events discusses how inputs to the FSM are filtered.

5.5.1.1 Locating Events

Some user interfaces have no need to locate events. If the interface is keyboard driven, all events will be uniquely associated with a key. Areas or objects on the display may be selected as a result, but these relationships are a programmed reaction to certain key sequences. A selected object, or the focus, may then be the first choice for receiving future events.

Locating a selected object becomes more complex when the event is associated with a particular screen location. Mouse events and pressing on a touch screen are two examples. In video games and animation, collision detection is performed by searching for objects that coincide with the location just moved to by another object. Starting with an event location, a search is performed to see which object or objects overlap with the event's location. It is trivial to write a function that checks to see if a particular area contains a particular point. When this function is called, you must ensure that the area and point are both relative to the same origin. If this check must

Listing 5.2 (continued)

```
/*
This function chooses which of the button's four images is
displayed, depending on the current state.
*/
void buttonDraw(Button * buttonPtr, int offsetX, int offsetY)
{
    switch (buttonPtr->state)
    {
    case UP:
        buttonDrawUp(buttonPtr, LIGHTGRAY, offsetX, offsetY);
        break;
    case PENDING_DOWN:
        buttonDrawDown(buttonPtr, WHITE, offsetX, offsetY);
        break;
    case DOWN:
        buttonDrawDown(buttonPtr, LIGHTGRAY, offsetX, offsetY);
        break;
    case PENDING_UP:
        buttonDrawUp(buttonPtr, WHITE, offsetX, offsetY);
        break;
    }
    buttonPtr->drawable.oldArea = buttonPtr->drawable.area;
}
```

be performed for every object on the display, it may become unacceptably slow. In this scenario, you can take advantage of the `Container` hierarchy. Start at the root `Container` and check its contents, descending into subcontainers only if they overlap the event location. In this way, you can avoid searching entire `Containers`. For this approach to work, it is important that none of the touch-sensitive objects extend outside the boundaries of their parent `Container`.

Listing 5.2 (continued)

```c
/*
This function draws the button in the up state. The color in the center
is passed in so that we can draw the button lit or unlit, depending on
whether we are in the PENDING_UP state or the UP state, respectively.
*/
void buttonDrawUp(Button * buttonPtr, int centerColor, int offsetX, int offsetY)
{
    int poly[12];
    int absoluteLeft = offsetX + buttonPtr->drawable.area.left;
    int absoluteTop = offsetY + buttonPtr->drawable.area.top;
    int absoluteRight = offsetX + buttonPtr->drawable.area.right;
    int absoluteBottom = offsetY + buttonPtr->drawable.area.bottom;
#ifdef USING_BGI
    setcolor(BLACK);
    setfillstyle(SOLID_FILL, WHITE);
    fillRectangle(absoluteLeft, absoluteTop, absoluteRight, absoluteBottom);
    poly[0] = absoluteLeft;
    poly[1] = absoluteBottom;
    poly[2] = absoluteRight;
    poly[3] = absoluteBottom;
    poly[4] = absoluteRight;
    poly[5] = absoluteTop;
    poly[6] = absoluteRight - BUTTON_BORDER;
    poly[7] = absoluteTop + BUTTON_BORDER;
    poly[8] = absoluteRight - BUTTON_BORDER;
    poly[9] = absoluteBottom - BUTTON_BORDER;
    poly[10] = absoluteLeft + BUTTON_BORDER;
    poly[11] = absoluteBottom - BUTTON_BORDER;
    setcolor(DARKGRAY);
    setfillstyle(SOLID_FILL, DARKGRAY);
    fillpoly(6, poly);
    setcolor(BLACK);
    line(absoluteLeft, absoluteTop,
        absoluteLeft+BUTTON_BORDER, absoluteTop+BUTTON_BORDER);
    setcolor(LIGHTGRAY);
    setfillstyle(SOLID_FILL, centerColor);
    fillRectangle(absoluteLeft+BUTTON_BORDER, absoluteTop+BUTTON_BORDER,
                absoluteRight-BUTTON_BORDER, absoluteBottom-BUTTON_BORDER);
    setcolor(BLACK);
    outtextxy(absoluteLeft+BUTTON_BORDER*3,
            absoluteTop+BUTTON_BORDER*3, buttonPtr->text);
#endif /* USING_BGI */
}
```

Figure 5.6 shows a display with four `Buttons`. When the event occurs, the first check made is whether the event is inside `Container` 1. Because it is not, `Buttons` 1 and 2 are not checked to see if they contain the event. Next, `Container` 2 is checked. It does contain the event, so `Buttons` 3 and 4, which are children of `Container` 2, are checked. In this case, neither contains the event, so no object is returned to the original caller. The important feature here is not just finding the correct `Button`, but avoiding many of the checks by using `Containers` wisely.

Listing 5.2 (continued)

```c
/*
This function draws the button in the down state. The color in the center
is passed in so that we can draw the button lit or unlit, depending on
whether we are in the PENDING_DOWN state or the DOWN state, respectively.
*/
void buttonDrawDown(Button * buttonPtr, int centerColor, int offsetX, int offsetY)
{
    int absoluteLeft = offsetX + buttonPtr->drawable.area.left;
    int absoluteTop = offsetY + buttonPtr->drawable.area.top;
    int absoluteRight = offsetX + buttonPtr->drawable.area.right;
    int absoluteBottom = offsetY + buttonPtr->drawable.area.bottom;
#ifdef USING_BGI
    setcolor(BLACK);
    setfillstyle(SOLID_FILL, BLACK);
    fillRectangle(absoluteLeft, absoluteTop, absoluteRight, absoluteBottom);
    setfillstyle(SOLID_FILL, centerColor);
    fillRectangle(absoluteLeft+BUTTON_BORDER, absoluteTop+BUTTON_BORDER,
                  absoluteRight-1, absoluteBottom-1);
    setcolor(BLACK);
    outtextxy(absoluteLeft+BUTTON_BORDER*4,
              absoluteTop+BUTTON_BORDER*4, buttonPtr->text);
#endif /* USING_BGI */
}
/*
This function erases the image of the button by drawing a rectangle, that
is the same size as the button, in the background color of the old parent.
*/
void buttonErase(Button * buttonPtr, int offsetX, int offsetY)
{
    int backgroundColor;
    int absoluteLeft = offsetX + buttonPtr->drawable.oldArea.left;
    int absoluteTop = offsetY + buttonPtr->drawable.oldArea.top;
    int absoluteRight = offsetX + buttonPtr->drawable.oldArea.right;
    int absoluteBottom=offsetY + buttonPtr->drawable.oldArea.bottom;
    backgroundColor = buttonPtr->drawable.oldParentPtr->drawable.color;
#ifdef USING_BGI
    setcolor(backgroundColor);
    setfillstyle(SOLID_FILL, backgroundColor);
    fillRectangle(absoluteLeft, absoluteTop, absoluteRight, absoluteBottom);
#endif
}
```

Graphics — 111

In the examples, I do not need to invoke events on any objects apart from the Buttons, so the containerLocateTarget() function checks the Buttons to see if they contain the event. It also checks Containers. If the event occurs inside a Container, the function is called recursively on that Container because it may contain Buttons. All other objects are ignored. It would be trivial to change this to make all objects locatable. This might be useful if you want to associate help with each object. The help text would become visible whenever you select the associated object. In other cases, you want simple shapes or bitmaps to be sensitive to allow direct manipulation of a diagram. Dragging a line that represents a section of a robotic arm could cause the motors to move the robotic arm to the new position.

When only one type of object is sensitive, you can keep a separate Container tree for those objects, or arrange the lists so that the sensitive objects are at the front of the chain of contained objects. Either of these approaches reduces the number of links followed to nonsensitive objects. The example code returns the first Button found and does not search any further. Overlapping Buttons are not supported.

containerLocateTarget() is the function that locates the object. Once the Button is found, the function buttonProcessFSM() is called to process the event.

```
Button *containerLocateTarget(Container *containerPtr, int x, int y)
{
    Drawable *iterator;
    Button *target;
    /*
    change the parameters passed in to local coordinates
    */
    int localX = x - containerPtr->drawable.area.left;
    int localY = y - containerPtr->drawable.area.top;
```

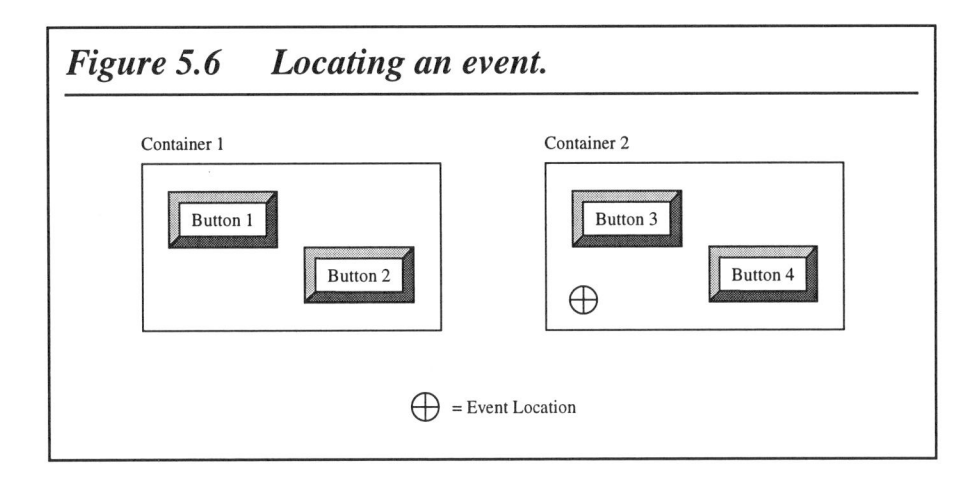

Figure 5.6   Locating an event.

```
    /*
    We iterate through the list of childern, checking for overlap
    on any containers or buttons.
    */
    for(iterator = containerPtr->containedListPtr; iterator != NULL;
                    iterator = iterator->nextContainedPtr)
    {
        if ( (localX >= iterator->area.left) &&
             (localX <= iterator->area.right) &&
             (localY >= iterator->area.top) &&
             (localY <= iterator->area.bottom) )
        {
            if (iterator->type == CONTAINER)
            {
                target = containerLocateTarget((Container*)iterator,
                                                    localX, localY);
                if (target != NULL)
                {
                    return target;
                }
            }
            if (iterator->type == BUTTON)
            {
                return (Button *)iterator;
            }
            /*
            All other types are simply ignored because they
            are not touch-sensitive.
            */
        }
    } /* end for */
    /*
    At the end of the loop, I found nothing to return NULL,
    so other containers will be checked. If this the
    top-level call in the recursion, the user did not
    hit a valid target.
    */
    return NULL;
}
```

In Chapter 4, I did not discuss the buttonMouseEvent() function, because it was beyond the scope of that chapter. I will look at it now because it performs an important job. All mouse events directed at the button have to be filtered to decide what inputs to the FSM will be generated. The containerLocateTarget() function shown earlier is used to locate the target of a mouse event. Once that target has been found, it remains the button in focus until it releases the focus. The events are passed to the buttonMouseEvent() function, which may release the focus by returning a NULL. By applying the rule that an object receives mouse events until it chooses to give up the mouse focus, I allow for shapes that can be dragged or resized by the mouse. However, I do not implement such shapes here.

The `drawableContains()` function (Listing 5.3) returns a `Boolean` to tell us if a `Drawable` contains a certain point. The `buttonMouseEvent()` function needs this because it must check that the mouse has left the `Button` in order to declare a `SlideOff` input to the FSM. Using the `containerLocateTarget()` function to check this would be more expensive because that function checks all `Drawables`, not just the one in which I am interested.

The transition tables used by the FSM are those shown in Chapter 4.

Listing 5.3 The functions used to track the location of mouse events.

```
Boolean drawableContains(Drawable *targetPtr, int absoluteX, int absoluteY)
{
    int localX;
    int localY;
    /*
    If the drawable has no visible parent,it cannot contain any point on the display.
    */
    if (targetPtr->parentPtr == NULL)
    {
        return FALSE;
    }
    if (targetPtr->parentPtr->visible == FALSE)
    {
        return FALSE;
    }
    localX = absoluteX - targetPtr->parentPtr->absoluteLeft;
    localY = absoluteY - targetPtr->parentPtr->absoluteTop;
    if ( (localX >= targetPtr->area.left) && (localX <= targetPtr->area.right) &&
         (localY >= targetPtr->area.top) && (localY <= targetPtr->area.bottom) )
    {
        return TRUE;
    }
    return FALSE;
}
/*
This function accepts the mouse event. It generates the appropriate input to the
FSM part of the button code. It is a filter that converts raw mouse actions into
FSM inputs. It returns the current button if this button still has the focus.
It returns NULL if it does not want to receive further events.
*/
Button *buttonMouseEvent(Button *currentButtonPtr,
                         int mouseX, int mouseY, int mouseButton)
{
    static int lastMouseButton = MOUSE_NONE;
    if ((lastMouseButton == MOUSE_NONE) && (mouseButton == MOUSE_LEFT))
    {
        buttonProcessFSM(currentButtonPtr, Press);
        lastMouseButton = MOUSE_LEFT;
        return currentButtonPtr;
    }
```

Listing 5.3 (continued)

```
    if ((lastMouseButton == MOUSE_LEFT) && (mouseButton == MOUSE_NONE))
    {
        if (drawableContains(&currentButtonPtr->drawable, mouseX, mouseY))
        {
            buttonProcessFSM(currentButtonPtr, Release);
        }
        else
        {
            buttonProcessFSM(currentButtonPtr, SlideOff);
        }
        lastMouseButton = MOUSE_NONE;
        return NULL;
    }
    if ((lastMouseButton == MOUSE_LEFT) && (mouseButton == MOUSE_LEFT))
    {
        if (drawableContains(&currentButtonPtr->drawable, mouseX, mouseY))
        {
            /*
            No FSM event occurs since the mouse is just moving around
            inside the button, with the mouse button down.
            */
            return currentButtonPtr;
        }
        else
        {
            buttonProcessFSM(currentButtonPtr, SlideOff);
            /*
            The mouse button still is down, but it is not active on
            any graphical button.  So we set it to NULL to start the
            sequence again the next time a button is acted upon.
            */
            lastMouseButton = NULL;
            return NULL;
        }
    }
    if ((lastMouseButton == MOUSE_NONE) && (mouseButton == MOUSE_NONE))
    {
        if (drawableContains(&currentButtonPtr->drawable, mouseX, mouseY))
        {
            /*
            No FSM event occurs since the mouse is just moving around
            inside of the button, with the mouse button up.
            */
            return currentButtonPtr;
        }
        else
        {
            /*
            No FSM event occurs since the mouse has moved out of the button,
            with the mouse button up.
            */
            return NULL;
        }
    }
    assert(FALSE);
}
```

The focus that manages the buttons is maintained as a `static` local variable in the `handleMouseEvent()` function in the `multi.c` module. This function is called every time a mouse button is pressed or when the mouse moves with the mouse button down. When it gets a `NULL` returned from `buttonMouseEvent()`, the function searches for another focus and passes the event to it immediately rather than wait for the next event. This allows for the case in which the mouse moves out of one button and into another in one event. The first button must process a `SlideOff` input, and the second button must process a `Press` input.

```
void handleMouseEvent(MouseEvent *mousePtr)
{
    static Button *mouseFocusPtr;
    Button *buttonActedOnPtr = NULL;
    if (mouseFocusPtr != NULL)
    {
        buttonActedOnPtr = mouseFocusPtr;
        mouseFocusPtr = buttonMouseEvent(mouseFocusPtr,
                    mousePtr->x, mousePtr->y, mousePtr->button);
    }
```

Listing 5.3 *(continued)*

```
/*
This function moves the FSM from one state to the next. Note that the FSM
itself is passed in, so the same function could be used for many Buttons
that behave differently, for example, some may latch and others may not.
*/
void buttonProcessFSM(Button * buttonPtr, ButtonInput input )
{
    int i;
    /*
    First loop through the FSM to find the appropriate transition.
    */
    for (i=0; GS_buttonFSM[buttonPtr->state][i].input != input; i++)
    {
        if (GS_buttonFSM[buttonPtr->state][i].input == NullInput)
        {
            /*
                The end of the list has been reached and no input matched the input.
            */
            assert (FALSE);
        }
    }
    /* Now i indexes the appropriate transition */
    /* Call the action function, passing the button, the input, and
       the nextState */
    GS_buttonFSM[buttonPtr->state][i].transitionFn(buttonPtr, input,
                GS_buttonFSM[buttonPtr->state][i].nextState);
    /* Change the state. */
    buttonPtr->state = GS_buttonFSM[buttonPtr->state][i].nextState;
    drawableDirty(&buttonPtr->drawable);
}
```

```
    /*
    Note that we cannot use an else here because buttonMouseEvent
    might have cleared the value of mouseFocusPtr.
    */
    if (mouseFocusPtr == NULL)
    {
        mouseFocusPtr = containerLocateTarget(GS_wholeScreenPtr,
                            mousePtr->x, mousePtr->y);
        /*
        If this is the button that we have just acted on, do not find
        it again. This covers the case in which the user releases
        the mouse button and the mouse is still inside the button.
        */
        if (mouseFocusPtr == buttonActedOnPtr)
        {
            mouseFocusPtr = NULL;
        }
        if (mouseFocusPtr != NULL)
        {
            mouseFocusPtr = buttonMouseEvent(mouseFocusPtr,
                        mousePtr->x, mousePtr->y, mousePtr->button);
        }
    }
}
```

5.5.2 Line Graphs

The Button is a more sophisticated input object than those discussed in Chapter 4 (e.g., keystrokes or menu selections). A more sophisticated output object is the line graph, or Trace, as I call it. This object merits discussion for two reasons: (1) it is common for embedded applications to display a real-time graph of a controlled or monitored parameter and (2) because it is a line graph and controlled by a continuous stream of data, the update algorithm may differ from that of an object receiving fewer events.

The most important decision is whether a buffer of the monitored data should be kept. With a buffer, the graph can be reproduced, which may be necessary if the graph is either scrolled or relocated on the screen. A buffer allows a draw function to reproduce the entire graph at once. The example implementation does not buffer the data, and thus, the graph can be drawn as the data arrives, but not afterward.

The graph scans from left to right. Once the right-hand side is reached, the graph restarts at the left-hand side. This is similar to the way nondigital oscilloscopes display data — one sweep at a time, each sweep erasing the last as it scans over it. Because the image is not scrolled, there is no need to redraw the entire graph a few pixels to the left each time a new sample is introduced. Word processors and video games have made scrolling windows a popular and accepted mechanism, but that does not mean scrolling is the only solution when the data is wider than the screen. On LCDs, moving objects tend to blur because of the settling time of the display elements. If an object moves fast enough, it disappears completely. This effect can be observed by rapidly moving the mouse on the LCD screen of a laptop computer. A scrolling line graph also suffers. Depending on the rate of movement, a scrolling line graph may blur or vanish completely. If you are prototyping on a cathode-ray tube (CRT) display, you may be shocked when you run your code on a target with an LCD screen.

Another problem with scrolling is that it can be time consuming to regenerate the entire graph and delete the old one. An alternative is to copy the area a few pixels to the left, but that involves drawing many pixels that were not affected by the Trace. The problem gets worse if there are grid lines on the graph that must be preserved as the Trace scrolls. The time taken to reproduce the graph may cause flicker. Ironically, the LCD is more forgiving here than higher quality displays. Because the old image takes some time to fade and the new image fades in slowly, the undesirable flicker effect that may be seen on a higher quality display will disappear.

A bar is drawn at the leading edge of the Trace, scanning over the image that was drawn on the last sweep. Although it looks neater to clear the trace completely when starting a new one, it gives the user less time to observe the first pass. In particular, the last few samples on the right-hand side are deleted very soon after being drawn, and the user might miss an important event (Figure 5.7).

The Trace object I will implement does not support drawing axes, labels, or other decorations. These items should be drawn around the trace object, preferably in the

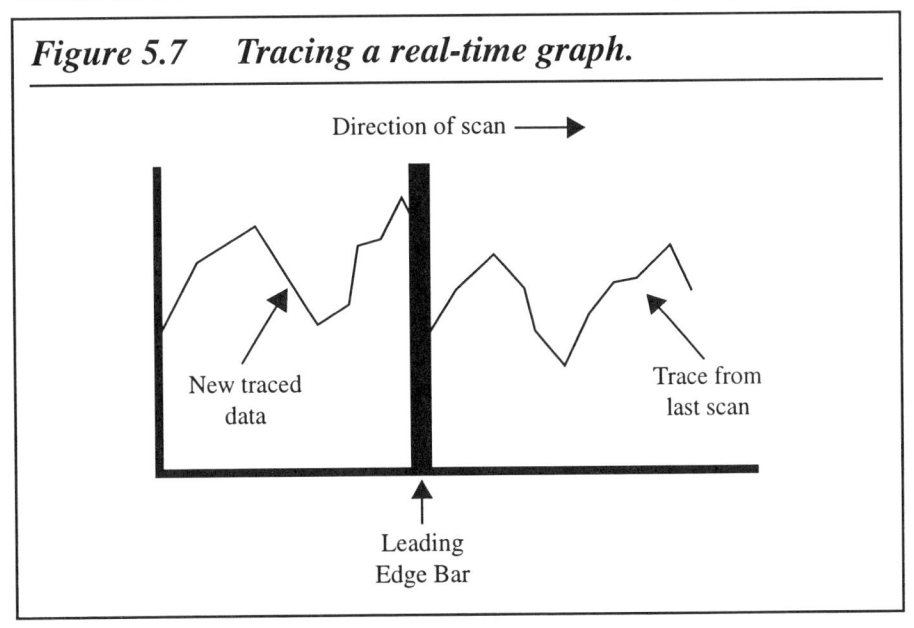

Figure 5.7   Tracing a real-time graph.

same parent container. Be careful not to overlap any of these objects with the graph; they will be erased as the Trace is drawn over them.

The traceCreate() function takes the number of pixels per sample as one of its arguments. This gives the Trace control over how far the leading edge moves for each new sample. In the implementation, this number must be an integer greater than zero. An alternative is to allow a fraction between zero and one, which permits many samples per pixel. The pixel displayed is then the average of the samples provided. However, this requires buffering of the data, so it is avoided in the example.

The traceSample() function is passed each data point as it becomes available. It displays it on the graph rather than deferring it to the displayRefresh() function. This breaks with the pattern used up until now, but completely redrawing the Trace object each time a new data point is added is inefficient. Another solution might be to make the Trace's draw function smart enough to know which parts of the trace to redraw. If the data points are buffered, this is my recommended approach.

Because the trace object does not measure time, the onus is on the application to call traceSample() at a constant frequency so that the graph scans smoothly. If your main event loop handles events that may take more than a second to process, you may have to consider getting a different task or an interrupt routine to call traceSample() to ensure that it is called in a timely manner. Separating these events from the event loop that controls the rest of the display can cause problems, especially if the drawing library is not thread safe. Most display hardware is not thread safe, so it is necessary at some level in the software to ensure that a lock is secured while drawing. This consumes CPU time and adds to the complexity of the code. If you are facing this choice, you may want to think about which actions cause the Trace to scan unevenly. If it is only a brief pause and occurs only when the user selects an uncommon feature, it may be acceptable. The user may not mind the graph jerking when he selects certain functions if it is reasonably smooth the rest of the time.

The example code uses a delay of one tick after each loop while polling for events. The tick is measured by the RTOS. There are a number of shortcomings to this mechanism, but even this simple scheme provides consistent speed across different processors. The trace performs one sweep in about five seconds, regardless of processor speed. This is what you want from an RTOS: predictable, consistent speed, not just raw processing power.

A delay of one tick means "delay until the next tick," so the delay could be from zero to one tick in length. The RTOS tick driven by the PC's clock has a frequency of 18.2Hz. If the PC had a clock with a higher frequency, the RTOS tick length would be shorter. If you delay for 10 ticks, the error of up to one tick would be less significant. The down side of a faster tick is that the processor spends more time processing those ticks for the RTOS.

Even if you ignore the error introduced by the resolution of the tick, the delay occurs after the polling work, so the actual interval between samples is the time to poll plus the length of the delay. The time to perform the poll may vary dramatically if a higher priority task becomes active while the poll is in progress. This inaccuracy may be fine for many systems, but if the user reads lengths of time from the screen, you may need to be more accurate. For this, synchronize with a constantly running timer that you need not stop and start each time you process a sample. Some RTOSs provide repeating timed semaphores, which are useful for driving events that must keep time consistently across a large number of periods.

Because the Trace may be updated often, the height is calculated at creation time to avoid some of the calculations when displaying a new sample. Small changes are required to the code that existed before the Trace was introduced. The TRACE type is added to the enumeration of DrawableType. Both the drawableDraw() and drawableErase() functions have new cases for the TRACE type. In both cases, they do nothing because the real drawing takes place when traceSample() is called. The extra cases have to be covered in the switch statement to avoid hitting the assert in the default case. Listing 5.4 shows the structures and functions that are added to our library to implement the Trace object.

Listing 5.4   The implementation of the Trace object.

```
struct traceStruct
{
    Drawable drawable;
/*
The absolute location is maintained to optimize drawing of the trace.
*/
    int height;
    int pixelsPerSample;
    int currentX;
    int currentY;
    int numberSamples;
};
typedef struct traceStruct Trace;
/*
This function clears the trace, ready for another sweep. Note that we do not reset
the current Y value because we may want to use that value to draw our first sample
*/
void traceReset(Trace *tracePtr)
{
    int absoluteLeft = tracePtr->drawable.parentPtr->absoluteLeft +
                       tracePtr->drawable.area.left;
    int absoluteTop  = tracePtr->drawable.parentPtr->absoluteTop +
                       tracePtr->drawable.area.top;
    /* erase the bar */
#ifdef USING_BGI
    setfillstyle(SOLID_FILL, tracePtr->drawable.parentPtr->drawable.color);
    setcolor(tracePtr->drawable.parentPtr->drawable.color);
    fillRectangle(absoluteLeft + tracePtr->currentX + 1, absoluteTop,
                  absoluteLeft + tracePtr->currentX + TRACE_BAR_WIDTH,
                  absoluteTop + tracePtr->height - 1);
#endif
    tracePtr->currentX = 0;
}
```

Because there is no buffer maintained in this Trace, there is a limit to how much functionality you may add. If you decide to add buffering, consider whether you need to store the data in the Trace object, or whether the buffers you may already have of the data, for other purposes, may be accessed from the Trace object. The biggest drawback of not having buffering is that you cannot redraw the data if it gets corrupted by another event. If you wish to implement a vertical line cursor that can move across the graph to mark points in time, you may be able to do it by drawing and erasing using the exclusive-or (XOR) function if the display hardware supports it. This effectively inverts the image that is covered, and you can invert it back again to recover the original, without accessing the data used to draw the original image. This technique is often used to implement a mouse pointer or drag-by-outline windows on desktop systems.

You can also increase this object's sophistication by making it understand the conversion between the real units and the number of pixels. If the object stores a unit conversion factor, the new samples could be multiplied by this factor as they arrive. This can be a big advantage if the trace is displayed in different sizes at different times. Unfortunately, a general scheme requires floating-point numbers, which you may wish to avoid for efficiency reasons.

Listing 5.4 (continued)

```
/*
This function takes one value and extends the trace line to the new value.
Only a Y value is passed in, and the X value is incremented by the same amount
every time. This means that this function must be called on a regular basis
for the trace to look smooth. The application layer must transform the real
value into a number of pixels before passing it in.
*/
void traceSample(Trace *tracePtr, int sampleHeight)
{
    int newX;
    int newY;
    int absoluteLeft;
    int absoluteTop;
#ifdef USING_BGI
    int mouseX, mouseY;
    Boolean hideMouse;
#endif /* USING_BGI */
    assert(tracePtr->drawable.parentPtr != NULL);
    assert(tracePtr->drawable.parentPtr->visible == TRUE);
    absoluteLeft = tracePtr->drawable.parentPtr->absoluteLeft +
                    tracePtr->drawable.area.left;
    absoluteTop  = tracePtr->drawable.parentPtr->absoluteTop +
                    tracePtr->drawable.area.top;
    /*
    The caller measures the height from the bottom (X-axis) of the graph.
    newY uses the top left as the origin because this is the origin for drawing.
    */
    newY = tracePtr->height - sampleHeight - 1;
    newX = tracePtr->currentX + tracePtr->pixelsPerSample;
```

5.6 Touch Screen Programming Techniques

The code shown in the previous section for locating an object applies equally to mice, touch screens, and any other screen pointer. If you use a touch screen, you should also consider the following factors.

The touch screen device driver should return events containing a location. There are no buttons that correspond to the mouse buttons, but the event in which the user releases his finger from the screen should initiate an action in response to the touch. In Chapter 4, I noted that keypads generally react to a button being pressed, and that there is no reaction to the release. On graphical systems controlled by mice or touch screens, the opposite is generally true. Although the initial press may have some visible feedback, actions are usually in response to the release.

When writing the touch screen device driver, you will find that the hardware signals require some interpretation. Generally, the screen consists of a grid of lines that

Listing 5.4 (continued)

```
#ifdef USING_BGI
    /*
    If the mouse pointer on the display overlaps the leading bar, it is necessary
    to hide the mouse and restore it after the display. The alternative is to always
    hide it for the update, but that causes a nasty flicker of the mouse pointer.
    */
    mouseRead(&mouseX, &mouseY);
    /*
    This complex if statement is checking whether any part of the mouse
    pointer overlaps any part of the leading bar.
    */
    if((mouseX > absoluteLeft + tracePtr->currentX-MOUSE_WIDTH) &&
       (mouseX < absoluteLeft + tracePtr->currentX +
               TRACE_BAR_WIDTH +  tracePtr->pixelsPerSample + 4) &&
       (mouseY > absoluteTop - MOUSE_HEIGHT) &&
       (mouseY < absoluteTop + tracePtr->height))
    {
        hideMouse = TRUE;
        mouseHide();
    }
    else
    {
        hideMouse = FALSE;
    }
#endif
    if (newX >= tracePtr->numberSamples)
    {
        /*
        We have reached the right-hand side, so clear and start over.
        */
        traceReset(tracePtr);
        newX = tracePtr->pixelsPerSample;
    }
```

may be light beams, sound waves, or one of several other technologies. The user's finger may break many of the horizontal and vertical beams. By averaging the vertical horizontal sets, a unique x, y-coordinate can be resolved. Note that this averaging can return a resolution that is double the resolution of the number of beams.

Typical touch applications do not use a visible cursor. If you require precise pointing you may need a visible hot-spot. The ideal cursor location is just above the finger, where the cursor is not obscured by another part of the user's hand. Changing the mouse cursor from the traditional northwest-pointing arrow is generally appropriate. An arrow pointing up or down or an icon of a hand all work well.

Touch-sensitive areas should activate on release. Once touched, the target object should visibly change. The change should include the borders to maximize the chance that the change can be seen around the user's finger. The user then has the opportunity to drag his finger away from the selected object. If the user lands on the wrong object initially, this allows him to cancel without completing the operation. Similarly, if he misses the object when his finger initially makes contact, the user can drag his finger until the target object changes appearance. The user then removes his finger to select the object.

Listing 5.4 (continued)

```
#ifdef USING_BGI
    /*
    Erase the old bar and draw the new one. Because these may overlap, I
    erase enough space to allow the new sample to be displayed. This reduces
    the flicker of removing and redrawing the overlapping part of the bar.
    */
    setcolor(tracePtr->drawable.parentPtr->drawable.color);
    setfillstyle(SOLID_FILL, tracePtr->drawable.parentPtr->drawable.color);
    /*
    The first argument to fillRectangle may add an extra 1 whenever currentX
    is non-zero. This is to ensure that we erase the first column, even
    though we never draw the bar in that position. It is important to
    erase whatever piece of the graph was visible in the first column.
    */
    fillRectangle(absoluteLeft + tracePtr->currentX + (tracePtr->currentX?1:0),
                  absoluteTop, absoluteLeft + newX,
                  absoluteTop + tracePtr->height - 1);
    /* add the new bar */
    setcolor(tracePtr->drawable.color);
    setfillstyle(SOLID_FILL, tracePtr->drawable.color);
    fillRectangle(absoluteLeft + newX + 1, absoluteTop,
                  absoluteLeft + newX + TRACE_BAR_WIDTH,
                  absoluteTop + tracePtr->height - 1);
    line(absoluteLeft + tracePtr->currentX, absoluteTop + tracePtr->currentY,
         absoluteLeft + newX, absoluteTop + newY);
    if (hideMouse)
    {
        mouseRestore();
    }
#endif
    tracePtr->currentX += tracePtr->pixelsPerSample;
    tracePtr->currentY = newY;
}
```

Lacking the ability to click on a selection, which you can do with a mouse, may not matter if you build the GUI from scratch. However, if some of the GUI is converted from a mouse-based system, you should carefully identify any interactions that may not map well. Pull-down menus do not work well, nor does any control that requires precise dragging. If a control requires an equivalent to a mouse button click, you may be able to use a pen-based system. Some resistive touch screens can detect the amount of pressure applied. The user's harder press on the screen might be interpreted as a click. However, clicking is not available on most touch screens. If you use a third-party toolkit for your controls, you should evaluate each control to decide if it is appropriate in your project and, if so, what the minimum size should be.

A finger moves slightly even if the user does not intend any motion. This causes a phenomenon called jitter. If the finger is on the border between two different x or y values, it may oscillate between the two. It is not an issue with a mouse because the user typically removes his hand from the mouse when he is not using it, allowing the mouse to be genuinely stationary. However, the lower resolution of a touch screen means that the size of the changes caused by jitter may be larger than the smallest movement possible on a mouse. Filtering the input values over time reduces jitter. Do not overuse filtering because it may cause the cursor to lag the finger. If you do not have a visible cursor and you do not drag objects on the display, the only place jitter will be visible is on the border of the sensitive area of a control, such as a button. As successive readings place the finger inside and then outside the button's sensitive area, the button's highlight (the visual indication that it is being touched) will turn on and off repeatedly.

If the button size is restricted by other information that must be displayed, you may want to increase the sensitive area so that it is larger than the button. If you do so, it is possible for two buttons that are close to each other to have overlapping sensitive areas. This works well if any touch event in the area of overlap is considered to be selecting neither of the buttons involved. Using this method also guarantees some dead space between the buttons. Figure 5.8 shows such buttons. Buttons 1 and 2 are so close, their sensitive areas overlap. The overlap area is treated the same as areas

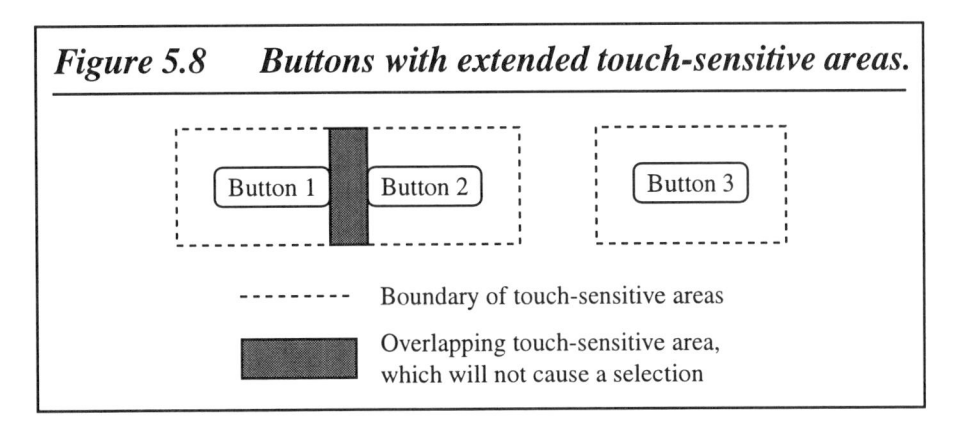

Figure 5.8   ***Buttons with extended touch-sensitive areas.***

containing no sensitive areas. Button 3 has plenty of room around it, and can use all of its sensitive area. There are no restrictions on placing nonsensitive graphics in the sensitive areas close to the buttons. Thus, you can display small buttons that are still easy to touch next to other information.

5.7 Multiple Dialogs

A simple GUI often consists of a small number of dialogs, or display configurations, that the user can switch between. These dialogs can include a "Setup screen," a "Standby screen," and a "Running screen." Each can be represented by a `Container`. A dialog requires data to allow it to be managed, as well as a `Container` to contain the shapes and controls that are visible to the user.

Each dialog is populated by adding a number of objects to the appropriate `Container`. As one dialog is entered, the previous dialog is removed from the root `Container` and the new one is added to the root `Container`.

Listing 5.5 *The function that builds up a dialog of two* Traces *and the two* Buttons *that can freeze those* Traces.

```
typedef struct
{
    Container *containerPtr;
    Trace *leftTracePtr;
    Trace *rightTracePtr;
    Boolean leftTraceLive;
    Boolean rightTraceLive;
    Button *leftButtonPtr;
    Button *rightButtonPtr;
} GraphsDialog;
static GraphsInfo GS_graphs;
void makeGraphs(void)
{
    Line *axisPtr;
    Text *textPtr;
    Text *instructionTextPtr;
    GS_graphs.containerPtr = containerCreate(140, 0, 639, 479);
    drawableSetColor(GET_DRAWABLE(GS_graphs.containerPtr), LIGHTBLUE);
    GS_graphs.leftTracePtr = traceCreate(30, 100, 230, 300, 2);
    containerAddTo(GS_graphs.containerPtr, GET_DRAWABLE(GS_graphs.leftTracePtr));
    GS_graphs.leftTraceLive = TRUE;
    /* add axis and labels */
    /* vertical axis */
    axisPtr = lineCreate(29, 100, 29, 300);
    containerAddTo(GS_graphs.containerPtr, GET_DRAWABLE(axisPtr));
    /* Add ticks to the axis */    axisPtr = lineCreate(25, 100, 29, 100);
    containerAddTo(GS_graphs.containerPtr, GET_DRAWABLE(axisPtr));
    axisPtr = lineCreate(25, 200, 29, 200);
    containerAddTo(GS_graphs.containerPtr, GET_DRAWABLE(axisPtr));
```

In some cases, pointers to objects contained in a Container are maintained so that they can be manipulated later. In other cases, once an object has been added to a Container, there is no longer any need to access it. If the Container is freed to allow reuse of its memory, it may be necessary to have a containerDestroyAll() function, which will iterate through its contents freeing each one.

Listing 5.5 shows the function that builds up a dialog of two Traces and the two Buttons that can freeze those Traces. Note how the pointers to the axis lines, which will not be manipulated later, are not stored, while pointers to the Buttons and Traces are stored in the GS_graphs structure. Through the GS_graphs structure, the Traces

Listing 5.5 (continued)

```
/* horizontal axis */
axisPtr = lineCreate(29, 301, 230, 301);
containerAddTo(GS_graphs.containerPtr, GET_DRAWABLE(axisPtr));
textPtr = textCreate(70, 315, "Time ->");
containerAddTo(GS_graphs.containerPtr, GET_DRAWABLE(textPtr));
/* Create the freeze button */
GS_graphs.leftButtonPtr = buttonCreate(50, 360, 150, 410,
            "Freeze", freezeButtonActivate, freezeButtonDeactivate);
containerAddTo(GS_graphs.containerPtr, GET_DRAWABLE(GS_graphs.leftButtonPtr));
/*
Now do the right-hand trace so that is similar.
*/
GS_graphs.rightTracePtr = traceCreate(270, 100, 470, 300, 2);
containerAddTo(GS_graphs.containerPtr, GET_DRAWABLE(GS_graphs.rightTracePtr));
GS_graphs.rightTraceLive = TRUE;
/* Add axis and labels */
/* vertical axis */
axisPtr = lineCreate(269, 100, 269, 300);
containerAddTo(GS_graphs.containerPtr, GET_DRAWABLE(axisPtr));
/* Add ticks to the axis */
axisPtr = lineCreate(265, 100, 269, 100);
containerAddTo(GS_graphs.containerPtr, GET_DRAWABLE(axisPtr));
axisPtr = lineCreate(265, 200, 269, 200);
containerAddTo(GS_graphs.containerPtr, GET_DRAWABLE(axisPtr));
/* horizontal axis */
axisPtr = lineCreate(269, 301, 470, 301);
containerAddTo(GS_graphs.containerPtr, GET_DRAWABLE(axisPtr));
textPtr = textCreate(310, 315, "Time ->");
containerAddTo(GS_graphs.containerPtr, GET_DRAWABLE(textPtr));
/* Create the freeze button */
GS_graphs.rightButtonPtr = buttonCreate(300, 360, 410, 410,
            "Freeze", freezeButtonActivate, freezeButtonDeactivate);
containerAddTo(GS_graphs.containerPtr, GET_DRAWABLE(GS_graphs.rightButtonPtr));
/*
Add text to describe the active keys.
*/
instructionTextPtr = textCreate( 150, 30, "B for Black graphs");
drawableSetColor((Drawable *)instructionTextPtr, LIGHTGREEN);
containerAddTo(GS_graphs.containerPtr, GET_DRAWABLE(instructionTextPtr));
instructionTextPtr = textCreate( 150, 60, "W for White graphs");
drawableSetColor((Drawable *)instructionTextPtr, LIGHTGREEN);
containerAddTo(GS_graphs.containerPtr, GET_DRAWABLE(instructionTextPtr));
}
```

can be accessed later to add sample data or to change their color. This `Container` is created and populated only once, although it may appear and disappear many times while the program runs.

A common design error is the creation of one event loop per dialog. This may seem simpler initially, but it is far harder to maintain. The same main event loop should be used for all dialogs. This allows events from sources not directly related to that dialog (such as off-screen keys and dials) to be handled in the same way, regardless of which dialog is currently displayed. Most behavior specific to a dialog is embodied in the callbacks from the controls the dialog contains. If part of the event loop must be managed differently in certain dialogs, the dialog should have enter and exit functions that can perform any initialization and tidy-up appropriate to that dialog. These functions may set and clear some flags that indicate how certain events should be managed.

The example program shows one key-handling function per dialog. If a dialog's `Container` is visible, its key-handling function is called. These key-handling functions return a Boolean to minimize the risk that a key is processed twice. In this way, keys can have different meanings at different stages. The events are still collected at the same point in all cases. The main event loop handles mouse and sample data events independent of which dialog is displayed. The event structure returned from `getNextEvent()` is a `union` capable of holding key events, mouse events, or sample data.

```
typedef struct
{
    EventType type;
    union
    {
        MouseEvent mouse;
        SampleEvent sample;
        char key;
    } data;
} Event;
void eventLoop(void)
{
    Event event;
    while(1)
    {
        getNextEvent(&event);
        switch(event.type)
        {
        case MOUSE:
            handleMouseEvent(&event.data.mouse);
            break;
```

```
        case KEYBOARD:
            handleKeyEvent(event.data.key);
            break;
        case SAMPLE:
            handleSampleEvent(&event.data.sample);
            break;
        default:
            assert(FALSE);
        }
        /*
        mouseHide() and mouseRestore() must be called because
        of the way DOS graphics manage the mouse. We do not
        want to draw over the mouse.
        */
        mouseHide();
        displayRefresh();
        mouseRestore();
    }
}
```

The dialogs described here play the same role as a window in Windows or X-Window. Unlike full windowing systems, these dialogs do not have frames or the ability to be moved around by the mouse. You can add such functionality by placing shapes inside the window to display a frame and by processing the mouse events appropriately. You can also create a frame object that is sensitive to mouse clicks and moves. Any actions on the frame will cause the parent `Container` to move. While such a mechanism is not difficult to implement, it is rarely appropriate in an embedded system. I discuss the use of windows further in Chapter 9.

5.8 Conclusion

This chapter supplies the tools for you to structure your graphical interface. More than any other type of interface, graphical interfaces require structured code. This is because without a reasonable amount of code reuse, the size of a GUI will bloat. You may ask if the techniques described in this chapter were used to implement the impressive video games that were available on home micros with memory below 64Kb back in the glory days of the Atari 800 and Commodore 64 — machines that correspond closely to the hardware in many modern embedded systems. Probably not. If high-speed graphics on minimal hardware is your goal, you will have to code everything in assembler, exploiting knowledge of the exact hardware configuration. Unfortunately, such code cannot be maintained, ported, or expanded easily. Early video games were written by lone cowboy programmers unconcerned with how easily their colleagues could understand the code they were writing. As the gaming profession matured and hardware became more capable, game writing evolved into a team activity and gaming engines hid the lower levels of hardware and software from the game designer. Your system requirements will determine whether you need the cowboy approach or industrial-strength coding.

Chapter 6

Systems Issues

Thus far, I have reviewed the techniques for managing the user interface and its interaction with the user. The user interface must also interact with the rest of the system if it is to be of any practical use. This chapter addresses a number of issues that extend beyond the user interface code to the rest of the system. The first is the number of processors in the system, and how the workload distribution affects the design. The second is safety issues, which by their nature have to be addressed at a system level. The third is the ability to retarget the user interface to other languages. This chapter gathers these otherwise unrelated issues and examines their possible impact on the user interface software structure.

6.1 Multiple Processors

As the cost of microcontrollers tumbles, systems with more than one processor are becoming more common. Multiple processors are almost mandatory in safety-critical systems but can also have cost and maintenance benefits for other systems. For example, using multiple processors in a car put the computing power where it is required. This is more cost effective than using copper wiring to route all controls to one CPU.

For the purposes of this section, the user interface processor is the processor that has direct access to the input and output hardware on the front panel. I will refer to the processor with which the user interface processor communicates as the slave processor, though it may be capable of performing a lot of its work independently. The slave processor may in practice be a collection of processors, but in this simplified discussion, I will refer to it in the singular. The user interface processor requests the slave processor to perform actions, but the slave processor can also request the user interface processor to perform actions, such as displaying some monitored values or alarm conditions. This division of work leaves the user interface processor free to manage the interface and little else. This often removes many of the hard real-time requirements from the user interface processor and places them elsewhere in the system. Thus, the amount of user activity should not affect the timing controls of other parts of the system, greatly simplifying the software design for both processors. In particular, the user interface processor may now have processor cycles available when the user is not interacting with it. The programmer is free to perform manipulations of the interface that would be too computationally intensive on a single processor design. This is particularly true of graphical systems that perform some type of system output animation. If you are displaying a graph of real-time data, the number of updates per second and the smoothness of scrolling depend on the amount of CPU time you can dedicate to the task.

It may be wise to provide visual/audio feedback to any slow user requests before they are completed. For example, if a key is pressed, the user interface processor should sound a key click and possibly display the button in a different way or light an associated LED. I will call this action the *initial feedback*, to distinguish it from the final result of the user's request. After the initial feedback, the user knows that the key has been acknowledged, even if the end result is not yet visible. It may be a second or more before the hum of the motor indicates that the real work is under way. The initial feedback should be performed soon enough to allay any concerns that the request was not detected. Immediate feedback is helpful regardless of the processor arrangement, but the approach is different depending on the number of processors involved. If there is only one processor involved, the feedback must not be computationally intensive because telling the user that the request is being processed actually delays the beginning of the real work. If there are two processors, the initial feedback could be delayed until after the request has been sent to the slave processor. The processing of the request and the feedback to the user can then be performed in parallel, and the amount of CPU resources consumed by the initial feedback will not be a critical issue.

Give early feedback for an input requiring more than a second to process.

If the user interface processor is idle while the second processor is working on a request and you know that the delay will be significant, then something more than a key click is useful to tell the user that the request is being processed. On a nongraphical system, you can use a flashing LED. If you know the length of the delay, a countdown timer or a bar graph is a good idea. An equivalent in the desktop world is the animated arrows you see in Windows 95 when a file transfer is taking place. If you want to animate something to show change, I suggest you find a way to confirm that part of the action has taken place between cycles of the animation. For example, in Windows 95, if a file transfer with a remote machine fails, the animation may continue running. The size of the data transferred does not change, so the user has some clue that all is not well. But this is only a small part of the dialog. Most of the dialog gives the impression that work is taking place. You cannot always check on progress without communicating with the slave processor (and thus slowing down its real work) but in some cases, the information is still available, either through an external sensor or because the slave processor sends constant updates for other reasons.

You should take a good look at your product to see what can be done with those idle CPU cycles. I once worked on designing a lung ventilator with a graphical front end. Early lung ventilators had bellows which the medical staff could see rise and fall, accordion style, reassuring them that the machine was performing. As the equipment evolved and started to use compressed gas, the bellows disappeared. I had to provide a screen on the GUI that displayed a bellows rising and falling. This would be used as a distance viewing screen when the operators were not near the machine. Animation was CPU-intensive, but because the processor dedicated to the user interface had no other responsibilities, it did not impinge on the breathing performance of the machine. In video games, the most obvious use of spare CPU cycles is to play music or insert sound effects.

There are other load-balancing issues between the user interface processor and the slave processor. Between the time a request is constructed on the user interface processor and the time it is executed on the slave processor, you will often have processing to carry out on either CPU. The request may have to be converted from real world units to engineering units. Motor speed may be entered as revolutions per minute, but the control software may prefer to use counts per millisecond. This conversion could be placed with the processor that has the most processing power to spare. Another common conversion is from fixed-point to floating-point arithmetic, and vice versa. However, avoid compromising the independence of the parts of the design. Another warning is that a conversion that leads to larger packets passed across the communications link is likely to make the communications more time consuming and, therefore, lose the benefit gained by moving the work.

6.1.1 Protocols

On a single CPU, the protocol between subsystems is defined in terms of function call prototypes or in terms of structures passed among tasks that use one of the mechanisms supplied by the RTOS. When communicating between processors, you may be able to mimic the local RTOS access if your vendor's RTOS supports multiprocessors.

6.1.1.1 Text Protocols

Many windowed tools have a command-line version, which has a GUI applied to it. The graphical front-end produces commands the user may have typed at the command line. This allows easy debugging at development time. If the commands are made visible to the user in the final product, he may better understand how the tool works. In some cases the user chooses to either operate the graphical control or type in the command. The user also can associate a command with a new button. Much of this power depends on the user's understanding of the command-line version.

A debugger that requires the user to enter a one-line command and then wait for a response fits this model. At the prompt for a text-driven debugger, you might type the command `evaluate motorSpeed`. The response is a number. On a GUI-driven debugger, double-clicking on the name `motorSpeed` in the displayed code generates the request. The response appears in a window containing the text `motorSpeed = 7`. The operations on the GUI generate the text requests of which the user need not be aware. Separating the user interface processor from the rest of the system can be useful, as the GUI can run on a processor separate from the rest of the application with a text protocol that defines the interface. The back-end of the application may not even be aware of whether the requests are coming from a GUI or from a person typing in the commands directly or, indeed, from a script file containing a previously prepared set of commands.

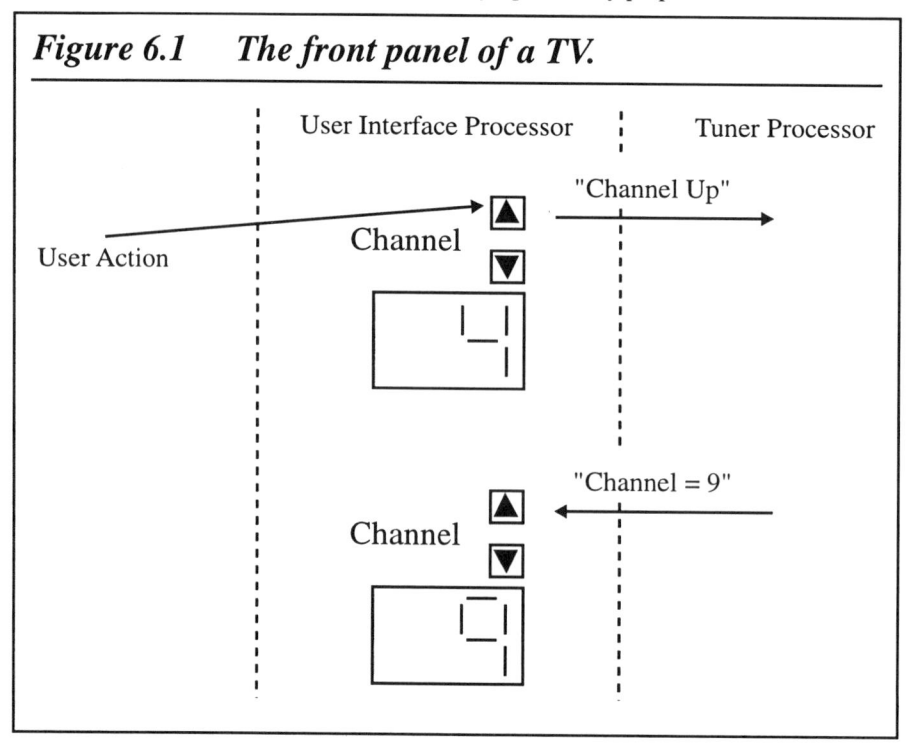

Figure 6.1   The front panel of a TV.

Embedded systems with several processors make good use of such an architecture. The example in Figure 6.1 shows the front panel of a TV. The channel is displayed by the user interface processor. But each time the channel is changed, the tuner, which acts as the slave processor, must decide which channel to display next. Only properly tuned channels are legal, so the user interface processor cannot simply increment the current channel number. Text messages passed between the two processors manage this interaction.

In testing, the tuner can be controlled from either a test system that passes strings entered by a test engineer or from a predefined script. All the interaction is humanly readable, making it possible to recognize mistakes without interpreting raw data. The user interface can be tested in a similar manner. After each user input, the text messages are checked to see if the appropriate request was generated.

The alternative to using a text protocol is to pass raw data across the link. Consider the packets used in the keyq example in Chapter 3, which are capable of passing key events or alarms.

```
typedef enum {KEY_PACKET, ALARM_PACKET, TIMEOUT_PACKET} PacketType;
typedef struct {
    PacketType type;
    union {
        KeyValue keyValue;
        AlarmStruct alarm;
    } data;
} QPacket;
```

This packet was passed on a queue in the keyq example but could just as easily be passed on a communications link. Unfortunately, the receiving end has to compile the same structure and the resulting memory layout has to be identical. This can be guaranteed only if both systems use the same compilers with the same compile-time flags on the same architecture. Otherwise, the size of the enumerated types and the amount of padding placed after fields of the structure may vary.

Alternatively, a text protocol can pass the key with the following statement

```
sprintf(buffer, "key = %d", keyValue);
```

The buffer is then passed to the communications link. At the receiving end, the keyValue is read into an integer of any size with the sscanf() function or a variant. The only restriction is that the range of keyValue can be expressed with the least number of bits set aside for that field by either processor.

The text protocol has avoided any issues of differently sized types or byte-ordering problems. This is a major advantage if a number of different processors are communicating. Even if the two processors in the embedded system are the same, you may wish to communicate with them from a test machine, which may have a different architecture.

If the interface between the user interface processor and the slave processor is a text protocol, there is a loss of efficiency. This loss is partly due to the number of characters passed between the two subsystems. Another contributor to the overhead is the parsing that must take place on the text, which has a more flexible format, in terms of spacing, than the equivalent binary format.

One project I was involved in maintained a text and a binary interface. The original intention was to use the text interface for debugging and replace it with the binary interface for the released product. The text interface was so much easier to manipulate that more powerful commands were added to the text interface alone. Thus, the text interface became faster than the binary interface, and eventually the binary interface was abandoned. I cannot claim that a text protocol is always better, but the advantages to development time can be so great that it is worth considering. For tightly coupled identical processors with tight real-time targets, a binary protocol is often necessary. For loosely coupled links between systems that communicate only occasionally and are developed by different groups, a text protocol proves itself worthwhile in the long run.

You can use tools such as Lex and Yacc (Levine et al., 1992) to define a grammar for the input if simple `printf()` and `scanf()` calls do not suffice. Hypertext markup language (HTML), the protocol used to define Internet Web pages, is probably the most popular text protocol in the world. HTML must be a text protocol because of the number of machines and platforms on which it is used. The fact that so many people have published personal Web pages demonstrates that this protocol is simple to use and is amenable to tool support.

If you must use a binary protocol, you may want to write tools, for test purposes, that convert between the binary protocol and a text format. Although you must freeze the system and look at the messages with a debugger, it is often advantageous to record a sequence of messages running at full speed and produce a printout later for analysis. Conversion tools also prove useful when generating test cases for one processor in isolation.

6.1.1.2 Stateless Protocols

If you want to allow one processor to reset without involving the other, then you must construct the protocol in such a way that each message is completely self-contained. It is usually straightforward to write the lower level communications software in such a way that it simply waits while the other side reboots and then retries the packet that was not acknowledged earlier.

It is difficult to write the higher level protocol so that it does not depend on the history of previous requests. This type of protocol is known as a stateless protocol. Consider an example of a protocol using state — one message tells which setting should change, and a second message contains the new value. If the receiving side reboots in between the two messages, it then receives a value. But the receiving side does not know how to apply it because the chosen setting was lost in the reboot. In a stateless protocol, such related information has to be contained in a single message. This leads to greater traffic on the communications link than would otherwise be the case. Consider a user interface that controls a bank of 10 motors. The user selects a motor and then performs a number of requests. The stateless protocol must indicate the motor in every command. A protocol that is not stateless can select a motor and use it as the default motor until the user interface indicates explicitly to switch to a different motor.

There are other approaches to this problem, such as storing state in some nonvolatile storage so that when one side reboots, it can restore itself to the state before the reboot. However, stateless protocols offer other advantages that I describe in the next section.

6.1.2 Multiple User Interfaces

Some systems have more than one user interface. The television with remote control is the most common example. One user interface is on the television itself, and the remote control provides a second way of accessing the same functionality. If you could buy a second remote control for the television, you would have three user interfaces. This works from a technical point of view but is a complete domestic disaster.

In some cases, the user interfaces access the slave processor simultaneously; in other cases, a primary user interface processor receives and processes requests from all the other user interfaces and only then passes requests to the slave processor.

Again, a stateless protocol offers advantages. It is feasible for the single slave processor to hold a separate state for each user interface. But this does not scale well, especially if it is possible that unforeseen user interfaces will be designed after the main system is shipped. In the other direction, if the user interface holds state about the slave processor, you must be able to notify each user interface when the state changes. Again, a stateless protocol avoids this issue entirely. In some cases, however, you may need to send information to each user interface if they are displaying information that is out of date.

If the user interfaces work differently from each other, the protocol has to support each one individually, or else be general enough to support anything that an interface may want to do. You may wish to imagine that you will use a dumb terminal to control the device when writing the protocol so that none of the packets defined are tied to something unique to the user interface.

Make communications interfaces generic and stateless to provide flexible future-proof systems.

You may also have a single user interface controlling several devices. For example, an alarm system can communicate with many sensing devices scattered throughout a building. The same logic applies to the protocol in these cases. If a sensor in one part of the building is reset, it should be able to send and receive requests after rebooting, without expecting the rest of the system to synchronize with it. This is possible as long as the protocol between sensors and the central user interface assumes nothing about previous messages.

The next level of complexity is to have multiple user interfaces and multiple devices to control — multiple television sets and multiple remote controls provide a slightly contrived example. In this case, the stateless protocol is necessary to avoid the matrix of relationships that otherwise must be maintained.

6.2 Safety

Safety is important for transport, medical, and aviation systems. Many industrial applications also have safety concerns, especially if dangerous chemicals or heavy equipment are controlled. Many of the same safety principles can be applied to systems in which no lives are at risk, but the financial risk is high. In cases of unexpected behavior due to malfunction or unforeseen external conditions, control of the device through its user interface may be limited. The user may also be in a situation in which the input to or output from the user interface cannot be trusted once the machine has malfunctioned.

Human error, often caused by inappropriate design of the human–computer interaction, is discussed further in Chapter 9. This section concerns itself with how to mitigate software errors and failures of components the software controls. Techniques of software-defect detection, such as testing and code inspection, form one layer of protection. Some failures, such as failure at the end of the life of a mechanical part, must be anticipated and accepted. The design should tolerate such failures or fail gracefully to a safe state.

Safety properties are often tied to the fundamental mechanical design, which dictates the ultimate limits of the system. Medical infusion pumps use gravity to produce flow through a small orifice. The orifice size and gravity itself dictate the maximum rate at which a drug can be administered. A software bug is unlikely to change the value of gravity, and although the orifice may be under software control, there is a fixed maximum opening. Such protection remains valid under user fault or software fault conditions.

Unfortunately, improving safety often reduces desirable system properties in such areas as cost, performance, and usability. For this reason, it is important that the person ultimately responsible for the safety of a project has a minimum investment in the other desirable properties. It is not trivial to maintain an organization in which this is the case. In small organizations, it is just not feasible to dedicate a person to the role of safety, although in some cases it can be combined with the testing function. The goals of safety are compatible with those of testing in that they both attempt to detect and eradicate any product shortcomings, and they may extend the launch date in return for better quality. Testing, however, tends to expose areas in which the implementation does not meet the specification. Safety engineering tries to ensure that the specification defines a safe system. Software bugs make spectacular headlines, and the public may perceive that implementation failures are predominantly caused by computer-related accidents. However, accidents caused by incorrect specifications outnumber accidents caused by software bugs (Leveson, 1995).

The user interface is the place where unsafe actions can most effectively be rejected and where unsafe states can be announced. If safety mechanisms are buried deep within the system, the user may have difficulty diagnosing the problem. At the time of an accident, quick and accurate diagnosis of the failure can be vital.

6.2.1 Safety versus Availability versus Reliability

Safety is the absence of accidents that cause damage or loss. Many systems remain safe simply by doing nothing. If an aircraft never moves, it is unlikely to cause harm, so a failure is not necessarily unsafe. Once committed beyond a certain point, the aircraft must continue to function in order to remain safe.

System availability is how often the system is usable. A design decision that enhances safety may reduce availability. If an airplane detects a ground temperature too low for a safe landing, its design may disallow take-off. This prevents the airplane from being used in a case in which the pilot knows he will land later in the day when the temperature is higher. It is a safe decision, however, because another failure may force the pilot to land immediately. In this case, there has been no failure because the system is performing according to specification. But system availability has suffered for the sake of safety.

Reliability is the absence of failures. Not all failures cause risks, and it is possible that many risks exist in a device even though the device has not failed to meet specification. For example, the system may have been used outside the parameters for which it was specified. This may represent operator failure, not device failure. Hence, reliability and safety are not synonymous. Software reliability is not measured in the same way as the reliability of mechanical and electrical components; software does not suffer from aging or fatigue. However, the frequency of software failures can still be measured and used to describe software reliability statistically. To build a safe system, you must allow for unreliable components in the two following ways. If the failure can be predicted, the operator is informed and the component replaced before a failure occurs. If the failure cannot be predicted but can be detected, then the system enters a fail-safe state.

The problem with software failures is that they can never be predicted. If they could be predicted, software engineers would simply remove them before releasing the software. They can sometimes be detected using the sanity checks described in the next section. The most problematic software failures are often those that are not recognized as failures by the system. If the user interface displays a monitored value that is wrong because of a software bug, but otherwise the system functions normally, the operator has no reason to suspect that the value is misleading. It is harder to design a safety net that prevents an error that can lead to an accident than to design a system that copes with components whose failure modes are well understood.

6.2.2 Sanity Checks

In your software, you have many opportunities to perform reasonability checks on the states of logical or arithmetic values within the system. If a sanity checks fails, the software knows there is a problem, though it may not know the exact cause. The check itself can be a Boolean expression placed inside the C `assert` macro, which is part of the standard C library (Maguire, 1993).

```
assert(x > 5);
```

This line indicates that the condition x greater than 5 must be true for the program to be valid. The standard C macro calls the `abort()` function to exit the program. In an embedded environment, you may wish to take some other action. The designer can choose such a failure to cause a restart in the hope that the restart resolves the problem. Many problems are transient, and the restart can ensure that all effects of the problem are removed. An alternative is to log only the fact that the sanity check failed, which makes sense for a system in which the restart is very expensive. This is true of telephone switches. AT&T uses software that searches through the many sanity-check failures that are logged to see if certain sequences occur or if failures repeat at certain frequencies. In more safety-oriented applications, it may be necessary to explain every sanity-check failure. Whether the sanity-check failures are presented to the user at the time they occur or whether they are logged for later analysis depends on the application. In the case of the telephone exchange, a voice synthesizer that communicates an error code to the user (the person making the phone call) is probably of no benefit.

Another example of a system that could not afford to restart while running was the computer on the Apollo 11 lander. It was capable of reporting error codes to the pilot if a sanity check failed. Neil Armstrong read the error code 1202 from a control panel as he made his final approach for the first moon landing. Mission control was able to deduce that this software problem would not abort the mission. Engineers at ground control knew that the error code was not critical unless the alert was continuous. Later investigation revealed that the CPU was running one extra task that took up processor time that should have been available for other tasks (Gray, 1992).

You may want to handle sanity checks of hardware or values that depend on hardware separately from sanity checks of the internal logic of the software. Hardware fails at the end of life, and the product should be tolerant of this expected event. Software defects can be found with internal checks, such as a string of, at most, 20 characters returning a length of 40. Such failures may need to be raised at a higher level in the field so that the designers are given the opportunity to investigate whether the software needs to be fixed.

On the user interface, sanity checks do not apply to user input. That is because the system should be specified to handle any user input and to give appropriate helpful feedback if the user attempts to enter illegal settings. A sanity check should detect this if the user interface receives data for display that is outside the set of reasonable values or if the data cannot physically be displayed on the available hardware. For example, you should never attempt to display a five-digit number on a four-digit display. The user may believe that the four visible digits are a valid value and act on this erroneous information. Some seven-segment display controllers give the choice between a character set that contains the letters A to F, to allow hexadecimal numbers to be displayed, and a character set that includes the letters H, E, L, and P (for example, the Maxim ICM7218). This allows one four-digit display to cry for HELP! More often, you also want a system-wide response to such a failure. If these letters are not available or if you are concerned about internationalization, the display could be left blank or display four minus signs. Avoid displaying zero in the fail-safe state, because zero may be a legitimate value and may be interpreted as such by the user. Values above the maximum allowed range should not be displayed as the maximum. Such clipping can also be interpreted as the currently measured value. This is a problem with needle gauges that go to maximum deflection when a greater-than-maximum current is applied.

You should choose limits for sanity checking output values carefully. If the device controls the temperature of a process up to 100 degrees, then the limit of the displayed user setting can be 100. However, the limit of the measured temperature should be much higher. The control process may have a tolerance, and the sensor used for measuring the temperature may also have a tolerance. This can lead to a monitored temperature above 100 degrees, which would still be a valid reading. External conditions can also drive the temperature higher. Do not interpret this as a failure of the sensor or controlling device, although an over-temperature alarm may be appropriate. Implying that the sensor has failed in such a case may lead to an inappropriate user response if the actual cause of the sanity-check failure is the overheating of the process.

In many applications, the safest response to out-of-range values is to alarm and allow the user to decide if it is due to a genuine anomaly in the controlled process or to a faulty sensor. This treatment may produce more alarms, but it could also mean that faulty sensors are replaced as soon as possible. Leveson (1995) points out that faulty sensors that are disabled or ignored are a major contributor to the severity of accidents because insufficient information is then available when you try to recover from an unsafe incident. The sensor is either known to be faulty and ignored or is not trusted because it was not maintained, so even accurate information may be ignored.

Set limits on the possible values manipulated in software to detect anomalies as soon as possible.

6.2.3 Fail-Safe State

Many systems enter a fail-safe state in response to system failure. Individual mechanical components can also be designed in this way. You can design a relay to fail open. If the relay fails, it does not allow energy to pass to a component capable of causing damage. On a keypad, the contacts should fail open, which means that the failed key cannot be used. This is preferable to the state in which the key is detected as permanently depressed, which may lead to unintended and unexpected actions. Software can identify a fail-closed key after it is detected as depressed for longer than a person would generally hold down a key. Unfortunately, the system response to those key presses may cause an accident by the time the key is detected as faulty. For this reason, some systems specify that repeated presses of the same key have no effect once the initial press is processed. This obviously has an impact on the design of the dialog between the user and the machine.

Sometimes, a component can be inserted to protect a subsystem. An electrical fuse may protect an electrical circuit from drawing enough energy to do harm. A valve that opens at a pressure limit can protect a pneumatic circuit from exceeding safe pressures. These components are not the source of failure, but they detect and mitigate failure, regardless of its cause.

Beyond individual component response, the system can have a fail-safe state. In many cases, you must simply remove power from components capable of creating a hazard and then cease all other operations. In other cases, you should generate an alarm to get immediate operator assistance. On some systems, a backup mode of operation is available. Often, the backup mode involves more direct control by the operator — for example, failure of an auto-pilot system.

For the user interface itself, you should ensure that when a display is frozen in a fail-safe state, the data does not mislead the user. Although a record of the state of the user interface at the time of failure may be useful in diagnosing the cause of the failure, the display should change dramatically enough that an operator can quickly tell that the device has either ceased to operate or is operating in a degraded mode. You may also need to display the steps required to leave the fail-safe state and re-enter normal operation. This is important when the fail-safe state is a rare occurrence, and there is only a small probability that the operator has seen it before.

If the fail-safe state is independent of the user interface processor, there may be an indicator light on the display, which is controlled by an electronic circuit with no software intervention. Such an indicator light is difficult to test. Be wary of providing an extra control line from software to control the fail-safe state indicator for test purposes. Such a line compromises the reliability of the circuit driving the light. A software fault might then lead to a state in which the fail-safe state indicator is on and the device is not in the fail-safe state. The correct way to test the fail-safe indicators is to force the device into the fail-safe state. This tests the indicator as well as other aspects of the fail-safe state.

> **Any safety device designed to protect against software failure should have no interactions with software, even for self-testing.**

In the fail-safe state, the user interface may attempt to display information to help the operator deduce the cause. During this time, you may wish to access information about the device. For this reason, the user interface should run in the fail-safe state, even if all mechanical controls are disabled.

During access to device data in the fail-safe state, you may need to ignore errors that normally lead to a response. For example, if the normal response to reading an out-of-range value from a sensor is to restart the system and you are already in the fail-safe state, you probably do not want to perform this action. The user may wish to observe the out-of-range value in order to establish that the sensor did, in fact, fail and how it failed. The user should also be able to access historical information stored in the device while in the fail-safe state.

6.2.4 Hazard Analysis

Several analytical techniques may be applied to an entire system in order to uncover and document potential hazards and the system's responses to them. The two most common techniques are fault tree analysis (Connolly, 1993) and failure mode effects analysis (Leveson, 1995).

6.2.4.1 Fault Tree Analysis (FTA)

FTA itemizes all hazards to which a system may be vulnerable. For each hazard, the events that can cause or contribute to that hazard are identified. FTA then identifies the secondary events that may lead to those events. This eventually leads to the primary event, such as a particular component failing at end of life. This process identifies events that may seem like minor faults but that can lead to an accident when combined with other factors. An over-temperature sensors failure may be harmless. However, the next time the temperature is too high, the alarm does not occur, which can lead to the temperature getting unacceptably high before an operator notices. The basic events can be assigned a probability used to calculate the probability of the hazard occurring because of any cause. Establishing probabilities for many events is difficult. It is usually more helpful to simply identify the paths leading to a hazard.

No fault should lead directly to a hazard. If all faults can be detected, then you assume they will be rectified before another fault occurs. Thus, many component failures or unacceptable events can occur, but if they are spread out in time or if the event ceases all activity, no hazard is reached

The user is a valid source of basic events. The user may create events via the system by operating the device, or the user event may not be directly detectable by the system. For example, if the system uses water, the user might accidentally add another fluid. The system may not detect that the wrong fluid was added, so mitigation of this accident has to be applied at a higher level. In the fault tree, the mitigation should be visible in a branch above the primary event.

If the event is an operation on the user interface, you can warn the user or require confirmation. If another event combines with this event on a path toward a hazard, you can give extra warnings to the user. If you can establish that the operation is never necessary when it might contribute to a hazard, you can disable the option on the user interface during those times. Raising the landing gear while an aircraft is still on the ground leads directly to an obvious hazard (it has really happened). It never makes sense to do this, so the option can be disabled.

Figure 6.2 shows a fault tree for a furnace explosion. From it, you can tell that the designers have sensibly installed a safety vent that opens if the furnace overheats. They have also implemented an over-temperature alarm. The operator can still generate an over-temperature condition by ignoring the alarm and turning on full power, which is designed for a cold start. You could design a timer that limits the length of time that the furnace may operate at full power. The designers can use this diagram to

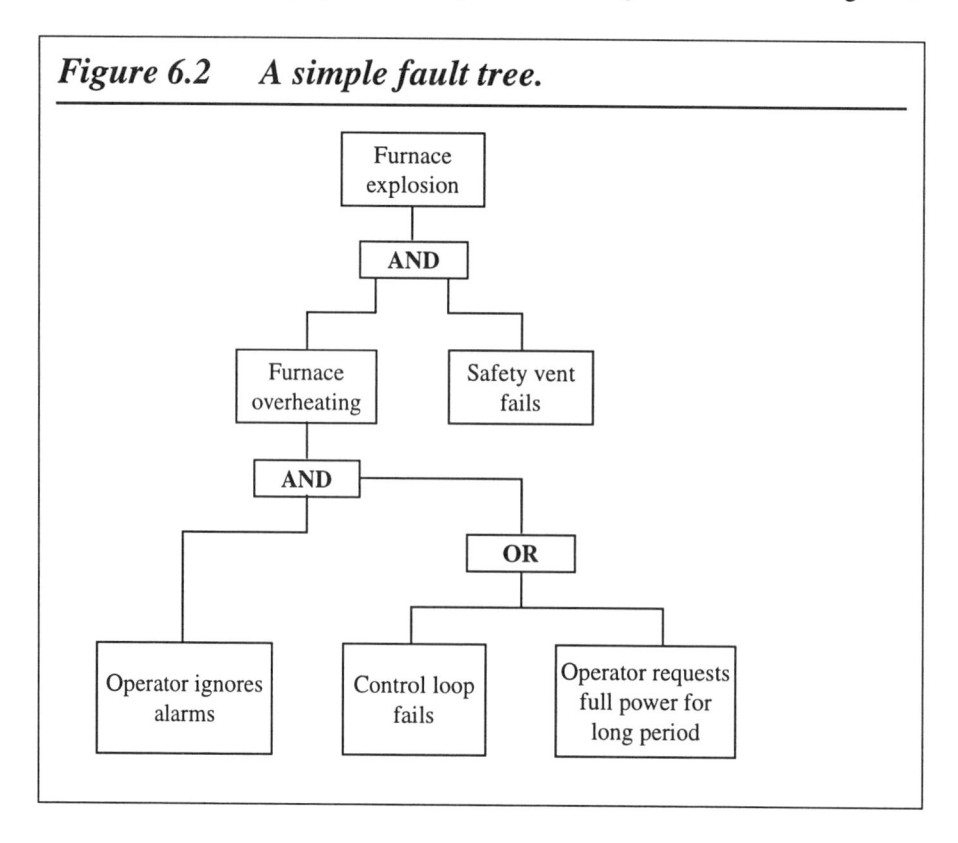

Figure 6.2   A simple fault tree.

claim that they have already taken enough precautions to prevent this user fault from leading to an accident. Other hazards, such as opening the furnace when it is hot enough to harm an employee, would form the root of another tree.

6.2.4.2 Failure Mode Effects Analysis (FMEA)

FMEA begins with a failure rather than a hazard. FMEA lists possible component failures and other undesirable events. It then lists the secondary effects of those events. Those events that lead to hazards help identify areas in which system safety is vulnerable. This technique can point out hazards not thought of when the hazards list was prepared for FTA. It is, therefore, useful to perform both types of analysis.

Events can be user-generated. For example, in air traffic control, you must consider the possibility of the operator fainting or being incapable of continuing as an operator, along with more ordinary operator mistakes. Sometimes, the system must protect itself against the user. The user of an automatic teller machine, for example, might deliberately attempt invalid transactions. The designer must assume that the user might try to subvert existing protections.

User events that do not involve input to the user interface may be impossible to prevent. However, the user interface can inform the user that the event was detected or can question him to establish that the correct procedure was followed. A common source of failure is service technicians who return a device to use without reconnecting all the cabling harnesses disconnected to perform the service. If appropriate signals are available, the user interface can warn the operator that this step was forgotten. Sometimes, a single line in the wiring harness is dedicated to detecting connectivity. For mechanical assemblies, or for other parts whose presence cannot be detected by the system itself, the system may question the operator to confirm that appropriate steps were performed.

6.2.5 Interlocks

An interlock is a mechanism that inhibits a certain sequence of events or allows only events in a certain sequence. Interlocks are commonly implemented in the mechanical or electrical system to prevent software from performing unsafe actions. For example, an in-service switch can indicate that a person is working on the machine. Software can detect this and refuse any requests to the front panel that may cause mechanical movement (and thus endanger the service technician). The switch can also open a relay that connects power to some or all of the mechanical system. Even if the software fails, mechanical movement is inhibited. Another example is a digital latch that can be turned on by software if the fail-safe state is entered. Operator intervention, such as power-cycling the device, is necessary to reset the latch. Software should regularly read the latch in normal operation to ensure that the fail-safe state has not been entered. This detects any unintended transitions within the software from the fail-safe state to the normal operating state.

Some safety professionals assume that it is impossible to predict the actions that software and the processors it runs on might take when a failure occurs. Therefore, when analyzing the system, you should assume that software may do anything that the hardware allows. It is often desirable to allow software to transition interlocks into the safe state but to allow transition to the nonsafe state only with an action guaranteed to have originated with the user, such as a directly wired switch or a power cycle. This prevents failed software from removing itself from the fail-safe state before appropriate system checks are performed.

The user should generally be unaware of the internal mechanisms that are in place to avoid hazards. However, any interlock requiring operator intervention must be justified to the user. Having to power cycle to get the device from one state to another may not be convenient, so documentation should explain that it is a safety feature. It might otherwise be perceived as a design deficiency.

You should give software read permission, but not necessarily write permission, on all interlocks. The software can then inform the user about why certain operations cannot be performed. You will also find it easier to detect and diagnose faults in the interlocks.

> **Guarantee that safety-critical interlocks can be reset only by direct operator action, to protect the system from errant software.**

As a system evolves, the software often becomes capable of controlling more of the system. Be wary of allowing some interlocks to become software controlled. In many modern devices, the on/off switch is a stateless button. This facilitates power management by the CPU, but it can have safety implications. A software fault may turn off the entire system, removing power from the backup alarms that should be the last line of defense against software failure. The Therac-25 radiation therapy machine, which massively overdosed several patients before it was removed from the market, was a successor to the Therac-6 and the Therac-20. The earlier machines did not use software control as extensively as the Therac-25. Replacement of hardware safeguards, present in earlier machines, with software checks led to a system vulnerable to software failures. The safeguards were removed to reduce cost and replaced with software checks, which have no recurring cost (Leveson, 1995).

6.2.6 Watchdogs

Software that is halted completely or executes so slowly that it cannot meet its real-time deadlines can be detected with a watchdog — an electronic counter that may be integrated into the microcontroller or stand as a separate digital device. The watchdog counts from a start value down to zero. If zero is reached, the device is deemed to have failed. This can result in the software resetting or entering a fail-safe state. Mission-critical software, in which no human intervention is feasible, must attempt to fully recover from the response to a watchdog failure. Stolper (1995) discusses responses to watchdog timeouts on the Mars Pathfinder spacecraft.

The watchdog timeout period is the time taken for the counter to reach zero. The software must reset the counter often enough so that it never reaches that point. The canine analogy states that a dog likes to be stroked and it keeps him passive. If you do not stroke him often enough, he will bite.

In many cases, the watchdog is restarted by writing a value or sequence of values to an address or sequence of addresses. A single address access is too simple and might accidentally be accessed by rogue software after the system suffers a software crash.

Usually, you can easily identify points in the software that should be visited more often than the watchdog timeout period. If the control loop examines the feedback from a sensor every 15ms, and the watchdog timeout period is 20ms, then the watchdog may be stroked every time the feedback loop begins. This ensures that the watchdog is reset with 5ms to spare. If the feedback loop is more than 5ms late, it is deemed to have failed and the watchdog will bite.

In user interface software, the problem is that user events cannot be guaranteed to occur at regular frequencies. Consider a system in which one task polls a keypad on a 10ms basis looking for key events. If a key is detected it is queued up for the main task to process. Because the main task spends most of its time waiting on events from the queue, there is no line in the main task that is executed often enough to stroke the watchdog. The polling task is guaranteed to run every 10ms, so this seems like a good point at which to stroke the dog. If the software stops running, the watchdog definitely bites. However, you should make this mechanism detect the widest possible family of errors. In the current scheme, if the main task cannot complete processing a key because it is stuck waiting for a resource or because it is in an infinite loop, the polling task still runs every 10ms stroking the dog at the appropriate interval. Several mechanisms can catch the locked main task problem. The polling task can perform several system sanity checks before allowing the dog to be stroked. The length of the queue to the main task can be checked. A queue-full condition can be considered a fault, because the main task has fallen behind further than expected. Alternatively, the polling task and the main task can take turns as the dog owner. Although the main task has work to do, it can stroke the dog every time a new message is taken off of the queue. When it has no work to do, it can pass the stroking responsibility to the polling task.

If additional tasks are involved, a more sophisticated mechanism may be necessary to check the state of all tasks before the dog can safely be stroked. The scheme depends on the hardware, as well as on the number of tasks and whether those tasks run regularly or in response to external events. Lowell (1992) describes a general-purpose scheme that can be used on a system running an RTOS.

6.2.7 Roles and Security

Security is closely related to safety. Many systems maintain their integrity under normal use, but must also protect themselves from abuse by users with limited privileges. Users with separate access rights are represented as separate roles within the software. You should pay careful attention to the transitions between roles. An automatic teller machine provides a good example. There is one role for the potential customer and one role for the account owner. The transition occurs when the identification number is accepted, not when it is entered. When a new feature is considered, you must decide which roles can access it. If you decide to print the time of day on the top-right of the screen, it could be displayed for all roles. If the account balance is to be printed in the same location, it is necessary that it is visible only after confirming the identification number. The transition time between the machine downloading the account information from the central database — so that the machine can display the balance — and the time when the role transition takes place is the period when most care should be taken.

In parts of the code where a role is necessary, it may be useful to check the role as a sanity check. This helps to track down paths through the software that lead from a nonprivileged function to a privileged function, without the appropriately controlled role change.

6.2.8 Separate Channels of Control and Monitoring

When control fails and the process behaves outside its specification, the failure must not disable all monitoring simultaneously. Consider a furnace in which the temperature is controlled by a feedback loop. One thermistor provides a feedback loop that controls the temperature to a target set by the user. The temperature measured by this thermistor is displayed to the user and is usually a couple of degrees higher or lower than the set value, depending on the accuracy of the components and on the control algorithm. Now assume that the thermistor fails and reads 150 degrees no matter what the furnace conditions. If the setting is 200 degrees, the control loop tries to compensate for the low temperature by increasing power to the heater. As the temperature rises, the control algorithm and the user both believe that the temperature has not changed. The control algorithm continues to increase the power to the heater. The user's natural reaction to seeing 150 degrees on the monitor is to turn the temperature control from 200 to a higher value to see if that increases the temperature. If the user tries to diagnose the problem at this point, a failed heater is seen as the likely cause. However, the user may perform some action, such as opening a furnace door, which would be safe at the displayed temperature but not at the actual temperature. The temperature is higher than the operator believes because the control loop is over-driving the heater while the interface displays an incorrect temperature.

A safer design is to use a separate thermistor for monitoring the temperature. It may or may not be possible for software to compare the two thermistors to detect failure, because they may be in different positions and may have different response times. However, the user is now in a situation in which either the setting or the monitored value is correct. They are both incorrect only if two components fail simultaneously. Once you decide to separate the monitoring and control channels, any common hardware or software in those two paths may compromise the independence of the paths. If the thermistors are humidity-compensated, then both paths are vulnerable to a fault in the humidity detector or a fault in the compensation algorithm.

In terms of user interface electronics, you can display several values on a single numeric display, with some control that allows the user to decide which of the multiplexed values he wishes to see. Figure 6.3 shows a user interface in which one four-digit numeric display can show any of four parameters. The LED connected to the top-right corner of the button indicates which of the four values is selected. When analyzing the system hardware faults, a failure of the numeric display now represents the loss of all those values. If you want to tolerate such a failure, it is better to independently display the settings and monitored values of the same system parameter. A disadvantage of the trend toward one large graphics display instead of a set of individual controls and displays is that the large display becomes a single point of failure, which removes all visibility of the system.

If one numeric display is used for the set and monitored value, a failure of half the display can cause confusion. If the two most significant digits of the display fail and the two least significant digits survive, the numbers displayed are in the range 0 to 99. So now it is possible for the user to set the value 150 while thinking he is setting the value 50. The measured value might be 149, assuming a small tolerance in the control loop. However, when the user looks at the monitored value, it reads 49. The most significant digit is missing again. Similar faults can occur if the decimal point segment fails. Although LCD displays are becoming more resilient to permanent failure, they are still

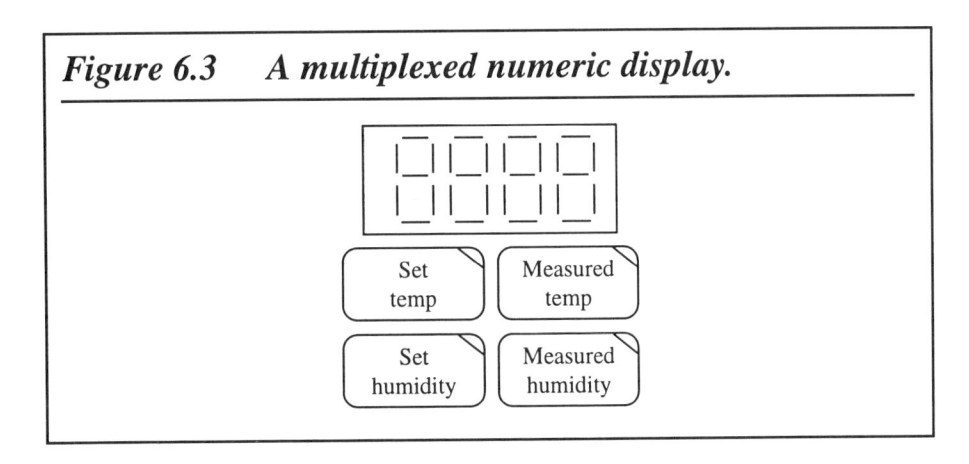

Figure 6.3   *A multiplexed numeric display.*

148 — Front Panel: Designing Software for Embedded User Interfaces

vulnerable to temporary corruption due to static electricity. Generally, this failure can be fixed by writing the correct value to the component again. Because the software has no way of detecting such corruption, it is a good preventive measure to do a refresh every few seconds on any LCD displays or other static-sensitive display elements.

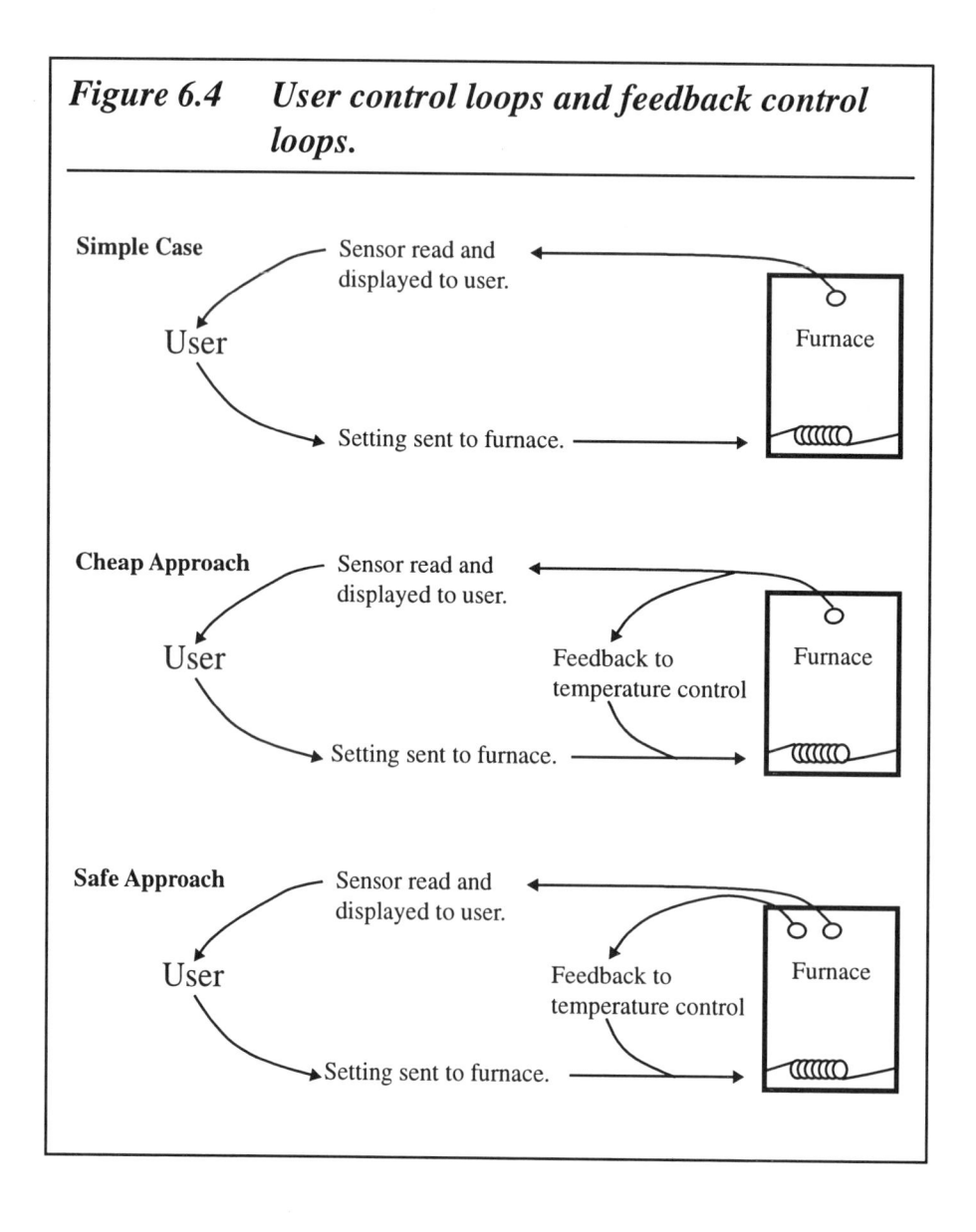

Figure 6.4   User control loops and feedback control loops.

You can reduce the severity of such faults if you separate the control channel from the monitoring channel as much as possible. The control channel begins with the user, and the monitoring channel ends with the user, thus forming a complete loop from the user to the process and back again. On some systems, this mandates two CPUs: one for monitoring and one for control. You may even wish to have two operators: one who manipulates the controls and one who monitors performance. You can see the behavior in the cockpit when a copilot reads instruments as the pilot performs a maneuver.

In the early stages of design, you can think of the system as two separate devices, one that controls the process and one that monitors it. Figure 6.4 shows the separate paths to and from a furnace. One path is to the heating element. The other path is from the temperature sensor. Ideally, the paths are separate, with no common components. In practice, another sensor is needed for control if you want a feedback loop that controls the heater. The cheaper approach is to reuse the same sensor. The safer approach is to place a new sensor in the furnace. This kind of redundancy can be more beneficial than simply duplicating components. When components are duplicated to do the same job, a mechanism must be applied to decide which component to trust under various circumstances. Such voting mechanisms can become more complex than the components they are trying to protect.

The user interface must annunciate alarms. If the alarms are based on a sensor reading, be sure to use the monitored value. You should consider the alarms as part of the systems monitor, not part of the controls, even if the user can set some alarm limits via the user interface. If a temperature alarm reads the sensor used to control the temperature rather than the sensor dedicated to monitoring, a fault in the control channel might disable the alarm. This is similar to the earlier scenario, in which the visual monitor and control algorithm depend on a single thermistor.

If the monitored values can be trivially compared with the settings to detect a fault, it may be tempting to automate the comparison and not burden the user with setting alarms or looking at the display to confirm acceptable performance. If you take this approach, the system becomes vulnerable to corruption of the settings. The displayed value should reflect the value in use rather than the last value entered. After a setting is entered, it can be convenient to simply leave that value on the display element. It is conceivable that the system might alter the value internally at some point, because of another external event or because a simple software defect corrupts the memory location where that value is stored. If that value is used for control and for setting the limits of a monitor, the user has no visibility of the current state. If the display is updated regularly based on the settings in use, such incidents may be detected by the user. This is especially important in systems in which settings may come from several sources or in which settings are temporarily changed internally during some procedures.

You may also compare the monitoring sensor with the feedback-loop sensor when the system is in a stable state. While the system changes temperature, the difference in the sensors' position within the furnace and the difference in response times may indicate slightly different temperatures with no fault present. It is better to check when the temperature is stable. This cross-check compensates for the reduced reliability associated with more components. If the probability of a single sensor is p, then the probability of a sensor failure when I have two sensors is 2*p.

Be wary of excessive averaging or filtering of monitored values. Although you must sometimes filter values to prevent narrow spikes from reaching the display, too much filtering hides genuine instability in the control loop. This can be a difficult trade-off if the user's perception of the device is not sophisticated enough to understand component tolerances and minor oscillations of a control loop. When the user sets a particular value, he may expect the monitored value to match that value exactly, without variation. Sometimes, filtering stabilizes the value enough to be acceptable to the novice user. The expert user, however, may not realize that some details are hidden. You can allow the user to see the unfiltered values when he is trying to diagnose a problem. Similar logic applies if the average is over several sensors rather than over time. Although the main display may show the average temperature of many sensors throughout the furnace, a more advanced mode may allow the user to query individual sensors.

6.2.9 *The Complexity Paradox*

Depending on the criticality of the system, you may employ some or all of the techniques outlined earlier. However, the safety aspects of your design must be simple. Failures can be difficult to generate in testing, and some are impossible to predict. Your system cannot exercise all the safety-critical code in all hazardous situations. Therefore, the probability of having defects in safety-critical code is higher than the probability of having defects in the corresponding amount of code in other parts of the system. It is easy to build a false level of confidence in a safety system that has run "trouble free" for a long time. Three possibilities exist: First the safety system is indeed stable; it neither raises false alarms nor misses genuine failures. Second, no system failures have occurred; you do not know if your safety net is operating. Third, the safety system is doing nothing, and dangerous situations are going unnoticed. Watchdogs, for example, are often disabled for debugging and then not reenabled. When the system misses hard deadlines, it may go undetected if the watchdog is your primary method for detecting such faults. The missed deadline may be propagated as occasionally incomplete or inconsistent data that can be difficult to diagnose.

When failure in hardware or software occurs during product testing, check that the safety net responded correctly. If not, fix the safety net first and then reproduce the failure to ensure that the safety net has been improved. Now you can be certain that the original fault is rectified.

Because finding defects in the safety net is difficult, you should make sure that there are very few. If a part of the safety net balloons into a large amount of code, step back and consider whether you added more hazards than you removed. When you must provide documented evidence of safety for regulatory approval, you may be very tempted to add many meaningless checks or to reduce safety by raising the chance of false alarms. Avoid checking for failures that have a vanishingly low probability of occurring unless the consequences of the failure are catastrophic. A few well thought out protection mechanisms work far better than many mechanisms that interfere with each other or compromise the overall design.

After the system is operational, you should conduct routine tests of the safety net to protect against component failure. Design failures should be removed before release of the system. The goal of these tests should be to confirm that components on which the safety net depends function correctly, especially those that are not exercised in the normal use of the device.

6.3 Translation and Internationalization

One of the requirements of any product is which languages the target platform supports. This requirement generally affects only the user interface developers and the documentation developers. It is not pleasant to near the end of a project and have to explain to your team that the results of their hard work can be released only to one market for the first few months because the translation process was not planned in advance. You might assume that translation means merely sending all the strings in the software and those printed on the front panel to a translation company that finds appropriate native speakers to do the translation. This is a simplistic view of a complex issue.

If you want native speakers who are also domain experts and are familiar with your industry and your family of products, then you may seek them within your own company or among your customers. For bulky work, such as manuals and help screens, you should consider having a professional do the translation and then having a domain expert review the work. The domain expert can ensure that the terms are consistent with those already in use on similar products and will match the conventions used within your industry.

You must also maintain consistency among the strings the software displays, the strings printed on the labeling, and the documentation's descriptions of the user interface. If a change is made in one of these places, the others must be checked for consistency. Once manuals are printed and hardware is manufactured, there may be pressure on the software to conform to the terminology used in the other two areas rather than the other way around.

It is a good idea to declare a freeze on the text strings at some point before the code is complete. This allows the translation process to occur in parallel with the final stages of bug fixing. It also allows the release of software for all languages simultaneously.

If several strings must be abbreviated to fit on the display, find abbreviations that are common to the industry. Avoid introducing ambiguity. For example, abbreviating "minimum" and "minute" to "min" can be confusing if the context does not make it obvious which one is intended.

6.3.1 Storing the Strings

C code expresses string literals in apostrophes (also called inverted commas). Ideally, you should place them in the code space, which means that they ultimately end up in read-only memory. You cannot be sure that your compiler does this automatically, so check the compiler's command-line options. Most compilers also have a switch that instructs it to search for identical strings and to merge them into one copy to save memory. If you use this technique, be careful that the logic of your code does not use the values of string pointers to choose different paths through the code, because two strings that are defined separately in the source code but that contain the same text are now located at the same memory address.

Usually you will confine all string literals to one code file and use pointers to access them from the rest of the code.

```
const char G_statusRunningString[] = "STATUS: Running";
```

The header file makes the string available to the rest of the system by declaring it

```
const char G_statusRunningString;
```

The `const` guarantees that any attempt to change the contents of the string pointed to by G_statusRunningString is flagged as an error.

```
G_statusRunningString[3] = 'x';
```

The preceding is an error.

You can use a `typedef` to give all your strings a fixed size. This can be useful if your display has a limited width. It is better to detect strings that are too long at compile time than at run time. I usually use something like the following.

```
typedef const char FixedString[20];
```

This ensures that when an array is declared as a `FixedString`, an error is flagged by the compiler if more than 20 characters are provided. It does not warn when there are fewer than 20, which means that your code should handle null-terminated strings shorter than 20 characters.

You can also declare all strings in an array and then access them using an enumerated type. I advise against this scheme unless you need to pass the index of the string across some communication link, where the index can be used to derive the string from an identical table. The enumeration method is difficult because inserting or removing one string can upset the ordering of the others. This increases the amount of recompiling required after a change. The enumerated type, which is in a header file, must also be synchronized with the array, which is declared in a .c file. There is no easy way to guarantee that they are both in the same order. With the pointers method described earlier, the order of the declarations is immaterial.

You should create different versions of the source file for each language. Errors may be introduced if the programmer editing the source for each language is not a native speaker. You can write a tool that assists with the substitutions, or you may be able to adapt a third-party tool.

At compile time, the makefile can choose the languages to build. You should decide whether to build all languages into one version of the software and allow the user to configure it, or to build separate versions of the software for each language. The size of the set of strings and the number of languages may dictate that only one will fit in the available ROM space. If there are language dependencies on the front panel, you may not want to allow the user to configure a set of French software with German labels on the front panel. The only time this might be appropriate is when the service technicians are predominantly English speaking and they might have to work with machines from many markets. If this is the case, it may be wise to allow two languages on each device: English and the language of the target market. If you opt to provide all languages, provide a means of reaching the language menu in a few steps, without reading the displayed strings. If the user accidentally changes to a foreign language, it may be frustrating to get back to his native language when he can no longer read the display.

If manufacturing personnel or service technicians must use equipment that is not in their native language, provide translation sheets. One copy of the translation should be in alphabetical order of the language implemented on the device. Another copy should be in alphabetical order of the user's language. The user can then find any string that may appear on the device. This scheme also allows the user to search for a string he has seen before on a version of the device in his own language.

You should also consider other localization issues, such as time and date formats and units of measure. These can be straightforward to implement in a configurable fashion if they are planned from the start. But they can be difficult to back-fit after the software is complete.

6.3.2 Character Sets

In the embedded world, you may not have the luxury the desktop world enjoys of choosing a font from the set supplied by the operating system. If you use a graphics module, there should be no technical limitation on the character set you wish to use. If you use a third-party toolkit, then it will probably be capable of reading fonts in one of the standard formats. Be warned, however, that the double-byte characters necessary to display Japanese and Chinese alphabets may require extra processing, and many toolkits do not support them. Many embedded systems require only the ability to display text, not the ability to edit it. This makes the task considerably easier. Whether a string is read left to right or right to left does not matter to the software.

But once the user has to type a line of text, such differences can make the software more challenging. *BYTE* magazine (Preston and Flohr, 1997) gives an overview of the issues you encounter when selling your product in the global marketplace.

Many simpler character displays, which contain one to four lines, are driven by the Hitachi HD44780 LCD module controller. Labrosse (1995) gives a detailed description of how to program this device, which has eight software-definable characters. A completely new mask can be made that allows all 255 characters to be redefined, but this has a tooling cost. The eight software-defined characters can be configured at run time. I know of one system in which the software is designed to figure out which characters are needed for a string as it is about to be displayed. The required characters are defined, and then the string is displayed. The character definitions are overwritten when a new string is displayed. Thus, the device can support the Polish and Russian alphabet without modifications to the English display module. The scheme depends on not having any strings that contain more than eight non-English characters, so it is fairly risky if your display has more than one line. It is not a good general-purpose solution, but it may allow you into some markets where the tooling cost would otherwise be prohibitive.

Once your character set becomes non-ASCII, it may not be possible to represent the string in your code in the normal way. Characters must use their hexadecimal representation. This has the disadvantage that the source code is no longer humanly readable. For example, say you want the French character ç, which is not usually available in the predefined fonts. We could define this letter to be pre-defined character number two. The string "garçon" now becomes: "gar\x02on". If a digit happened to follow the hexadecimal number, for example "ç7 abc" then you would have to separate the two as follows: "\x02""7 abc". Ending a string, and beginning a new one will cause the strings to be concatenated by the preprocessor in C, but the interpretation of the hex number will be restricted to \x02. If the string "x\027 abc" had been used then the compiler would have interpreted \x027 as a single byte.

If you need a custom mask to fit your needs, then design the custom mask to fit all the languages you intend to support. That part can then be used on all displays, avoiding issues of inventorying the variations separately.

6.3.3 Localizing Hardware

You can design hardware to avoid language issues. This has benefits in terms of the inventory of languages that must be stored. Manufacturing the correct mix of languages to fit the demand in various markets is a delicate balancing act, so try to avoid the issue entirely. The following are several schemes that may help you.

If you put the entire user interface on a graphical display, all translation can be handled from software. If there are hard keys or titles, you may be able to supply attachable labels that are applied at installation or by the user. All units can include labels for all languages. If you follow this approach, be sure the labels are sturdy enough to withstand long-term use and cleaning. You may be able to protect them with a clear plastic slider.

You can also use soft keys, which have no printed labels but are adjacent to a screen that displays software-generated text associated with each key. This technique is popular on automatic teller machines. However, this approach is a major design decision that should not be taken on the grounds of translation alone.

At first, icons may seem to be an attractive option. However, few industries have a well-known set of symbols that are used across all markets. If you can find appropriate icons, you may ease your internationalization problem. But make certain that you have not replaced the internationalization problem with a more severe usability problem. The icons are likely to be of real benefit only if you can use them to replace all text. On most products, this is unlikely. Once you translate 10 buttons, translating 20 is no more difficult. I examine the use of icons in Chapter 9.

If the device contains custom LCDs that contain words, then a separate mask may have to be created for each language. You may also have different per-unit costs, depending on the volumes required for each version. If the LCD is backlit, consider the following as a cheaper alternative. Put a solid block on the custom mask in the place you want the word to appear. Place sticker, or overlay, in the same shade of gray as the LCD, over the solid block. The sticker must have a translucent portion forming the word you want to display. While the block is turned on, it appears as a solid block on the LCD. When the block is turned off, the backlight shines through the overlay, making the word visible. Different languages require different overlays, but the overlay costs may be significantly less than custom mask costs. The number and size of your markets will help you decide whether this approach makes sense.

Chapter 7

C++ for Embedded Systems

You should consider C++ for several reasons. The current generation of embedded products have a more sophisticated user interface than previously, and C++ is the language of choice for desktop user interface software. But how useful is C++ to the user interface? The answer influences your decision to use C++ for part or all of the system. Several issues should be addressed before a team chooses the C++ route. You must address the choice of tools and the performance issues. The issues are not just technical. Educating the team may require a larger budget than purchasing new compilers.

This debate produces no outright winner. You as the engineer need a measure of the appropriateness of C++. This chapter attempts to supply you the information needed to make that judgment. If C++ is chosen for an embedded user interface, there are several features of the language that will be important to you, such as the means to place data in certain locations in memory and mechanisms to reduce the C++ performance overhead. These features are discussed in greater detail than is the rest of the language, which is too large to cover in a single chapter.

The first section of this chapter is an introduction. Skip it if you already have experience with C++. This chapter does not try to teach all of C++. Many volumes have been written on that topic, and it is neither appropriate nor economical to reproduce that information here. I will provide you with enough information to make it possible for you to decide whether C++ is appropriate for your work. The decision to change from C to C++ is a complex, risky one. If you choose to make the transition, you should know both the costs and the benefits.

Because of the large range of features C++ offers, several features are ignored in the discussions and not used in the example code, including the iostreams library and

157

passing arguments by reference. These features involve learning a lot of new syntax. They are not features central to the language and should not effect your decision to choose the language.

There are many complete introductions to the language, including Eckel (1995), Stroustrup (1991), and Ellis and Stroustrup (1990). Meyers (1996) and Cargill (1992) may be of interest to those looking for information on the style and philosophy of C++ programming.

7.1 Introduction to C++

This introduction presents an overview of C++. I explain the major features in enough detail so you can see why they are applied and what kinds of design they support.

Although C++ supports object-oriented programming, it does not enforce it. Any program that can be written in C can still be written in the same way in C++. The extensions that C++ provides enable the programmer to define and create objects and to facilitate code reuse. When you program in C++, do not feel obliged to use all the following features at every possible opportunity. At first, use them only where it is patently obvious that the C++ feature provides a better solution. As you become more of an expert, you will recognize good uses of these features more readily.

The principles of encapsulation and polymorphism were described in Chapter 2. You may wish to review those sections now in order to understand the rest of this introduction better.

7.1.1 Classes

The class is the most basic building block of C++. A class is a combination of a structure and the functions that manipulate that structure. These functions are referred to as *member functions*. Each data member and member function can be described as either public, meaning that any part of the program can access it, or private, meaning that only member functions can access it.

Once a class is defined, any number of instances of that class can be created. This is exactly the same as defining a structure and declaring an instance of the structure. Any instance of a structure may be global, local, or allocated on the heap in C. Any instance of a class may be global, local, or allocated on the heap in C++. In C++, instances are also called *objects*.

In C, the following code and data structure could manipulate stock in a shop.

```c
struct productRecordStruct {
    int numberOfUnits;
    int unitPrice;
};
typedef struct productRecordStruct ProductRecord;
getStockValue(ProductRecord *prPtr)
{
    return prPtr->numberOfUnits * prPtr->unitPrice;
}
void setUnitPrice(ProductRecord *prPtr, int newPrice)
{
    prPtr->unitPrice = newPrice;
}
```

You want to present the user with a function to increase or decrease the number of units because the rest of the software wants to manipulate the number of units this way rather than explicitly change the value. Either a delivery arrives, increasing the number of units, or a sale is made, decreasing the number. An increase and decrease function makes the program less prone to error because the caller does not manipulate the `numberOfUnits` field directly.

```c
void inreaseNumberOfUnits(ProductRecord *prPtr, int units)
{
    prPtr->numberOfUnits += units;
}
void decreaseNumberOfUnits(ProductRecord *prPtr, int units)
{
    prPtr->numberOfUnits -= units;
}
```

To provide better encapsulation, you may decide that this structure should be manipulated by these functions and no other. This can be enforced in C by not allowing the programmer any visibility of the structure itself. The following header file and a library or object file that contains a compiled version of the previous code would achieve this.

```c
struct productRecordStruct;
typedef struct productRecordStruct ProductRecord;
getStockValue(ProductRecord *prPtr);
void setUnitPrice(ProductRecord *prPtr, int newPrice);
void inreaseNumberOfUnits(ProductRecord *prPtr, int units);
void decreaseNumberOfUnits(ProductRecord *prPtr, int units);
```

When you use this file, you cannot see the contents of the structure or manipulate them directly. This is a crude form of encapsulation since it depends on the separation of files within the program. This may be a good match for large structures, but not for small ones.

When the structure is not visible in a compilation unit, the size of the structure is not known and instances of the structure cannot be declared. The supplied library now must provide a function allowing the user to create instances of the structure, either on the heap or in a static space allocated by the supplied module. This seriously restricts the way the structure can be used. If you want to let the calling functions declare instances of the structure on the stack, then the structure must be exposed.

Another problem is that the code is cumbersome. The pointer to the structure must be passed to each function, and that structure is repeatedly referred to from within those functions. Following is a C++ class that performs the same task.

```
class ProductRecord {
private:
    int numberOfUnits;
    int unitPrice;
public:
    int getStockValue(void);
    void setUnitPrice(int newPrice);
    void increaseNumberOfUnits(int units);
    void decreaseNumberOfUnits(int units);
};
```

The class definition is similar to a structure definition except that it also contains the function prototypes. The data or the member functions can be in the public or private section of the class, depending on the visibility required. In this case, all the data is private. This is typical because public data defeats the encapsulation you are trying to achieve. This class declaration may appear in a header file, but even though you can physically read the definitions of the private data, the compiler does not allow you to access it.

The functions are implemented as follows.

```
int ProductRecord::getStockValue(void)
{
    return numberOfUnits * unitPrice;
}
void ProductRecord::setUnitPrice(int newPrice)
{
    unitPrice = newPrice;
}
```

```
void ProductRecord::increaseNumberOfUnits(int units)
{
    numberOfUnits += units;
}
void ProductRecord::decreaseNumberOfUnits(int units)
{
    numberOfUnits -= units;
}
```

Note that the structure itself is no longer passed as a parameter — it is assumed. When the name of a member of the class is required, it can be accessed using the name alone, without the preceding `pr->` used in C code. If the class instance, or object, is not passed as a parameter, then how does the function know which object it should use for a given call? To answer this, look at the following to see how a program can use this class.

```
int value;
ProductRecord pr;
pr.setUnitPrice(10);
pr.increaseNumberOfUnits(100);
value = pr.getStockValue();
```

The member functions can be accessed with the same syntax used to access a member of a structure. In this case, dot syntax is used. When you have a pointer to an object, the `->` syntax is used. In this way, each function call is associated with a unique object, and that particular object is accessed in the member functions. In practice, a pointer to the object is placed on the stack along with the other parameters, but you do not need to be aware of this hidden parameter.

This technique of closely linking each function with a set of data has many advantages. The set of data that one function can manipulate is limited. If another object needs to be accessed, control is passed to a member function of that object. This arrangement gives the program better structure. The adage, "Good fences make good neighbors" holds for software as well as gardens. If each part of the program operates on its own turf and does not interfere with data belonging to another class, the program is easier to understand and maintain. Note that C allows the grouping of data and code, and that is what I have sought in the other chapters of this book. For the curious, Schmidt (1997) explores the similarities between well-structured C and the features built into C++.

Now that you have provided this protection, you can no longer directly access the data in the structure when you want to make a simple read-access to the data. Thus, for a data member x you often create a `getx()` function, called an *access function*. You may also have a `setx()` function if you want the caller to be able to directly update the value. By choosing to provide one or both of these functions, the class designer is choosing to provide read access, write access, or both to the data member. Just as with a file system, restricting privileges facilitates a more secure system. Hiding each data member behind a set of function calls also allows the class designer to later change the storage representation of x without changing the public interface.

Classes give the programmer a place to put things. If a constant is required within the class but not elsewhere, then it may be declared within the class scope. Similarly, if a type is defined privately within the class, it is available to the class' member functions, but not to the rest of the program. For example:

```
class ProductRecord {
private:
    typedef int ProductId;
    static const int maxUnitPrice;
    ProductId id;
    int numberOfUnits;
    int unitPrice;
public:
    // as before
};
const int ProductRecord::maxUnitPrice = 500;
```

Even though the `static const int` is declared within the class, it will not take up space in each instance of the class. It is actually given a value outside the class definition, and that value is the same for all objects of this type.

The `typedef` has no more and no less overhead than a `typedef` in the global space, but its visibility is restricted. If the type is declared in the public portion of the class, then it can be accessed. But it must be referred to as `ProductRecord::ProductId` if it is accessed from outside the class. The name of the type is thus protected against clashing with the name of any symbol from another class. Similarly, a member function name may be the same as the name of a member function of another class, but the class scope applied to the names resolves any potential ambiguity.

A class can also have nested classes or structures. In general, the class provides anything you want, to allow the class to be used while keeping it hidden from the rest of the program.

7.1.1.1 Constructors and Destructors

After you group the data and code, you want an initialization function to put the data into a reasonable initial state. C++ formalizes this idea with *constructors*. A constructor is a function that runs when an object is created. It lacks a return value, and it has the same name as the class. You can have many constructors, each taking different numbers or types of arguments. The following adds two constructors to the `ProductRecord` class.

```cpp
class ProductRecord {
private:
    int numberOfUnits;
    int unitPrice;
public:
    /* The first constructor takes no arguments */
    ProductRecord(void);
    /*
    The second constructor allows us to initialize the private data
    */
    ProductRecord(int initialPrice);
    int getStockValue(void);
    void setUnitPrice(int newPrice);
    void increaseNumberOfUnits(int units);
    void decreaseNumberOfUnits(int units);
};
ProductRecord::ProductRecord(void)
{
    numberOfUnits = 0;
    unitPrice = 0;
}
ProductRecord::ProductRecord(int initialPrice)
{
    numberOfUnits = 0;
    unitPrice = initialPrice;
}
```

The compiler guarantees that one of these functions will run when the object is created. Note that one constructor cannot call the other. The following two declarations each declare a ProductRecord on the stack.

```cpp
// This will call the first constructor with no arguments
ProductRecord pr1;
// This will call the second constructor with one argument
ProductRecord pr2(10);
```

A declaration with a different number of arguments and which has not been defined for the class is a syntax error. Note that the initialization arguments are not necessarily values stored in the data fields. You can use those arguments as you wish.

The constructors shown here are simple. In some cases, memory or other resources are allocated in the constructor. At some point, the resources must be freed, and you should minimize the risk that you might forget to do so explicitly. For this reason, the *destructor* was invented. It is called when an object goes out of scope — that is, at the end of the function that declared the object or when the object is deleted from the heap. You will learn more about how to create and destroy objects on the heap later. The name of the destructor is the class name preceded by a tilde (~), so if ProductRecord has a destructor, it is defined as follows.

```
ProductRecord::~ProductRecord(void)
{
    /* do some tidy up */
}
```

Objects on the heap need special consideration. The malloc() function allocates a block of memory but does not have any type information, so it cannot run the constructor for us. C++ includes an operator called new, which allocates space on the heap, similar to malloc(). Because new knows the type of the object, it runs the constructor. You can also pass any constructor arguments required.

```
ProductRecord *prPtr1 = new ProductRecord;
ProductRecord *prPtr2 = new ProductRecord(10);
```

The argument passed in to the new operator is passed on to the constructor. You do not need to pass the size required as you would with malloc(). That information is derived from the type name. This removes a significant source of errors with malloc(), with which it was possible to allocate a space of the wrong size for a structure.

These objects now live on the heap and can be freed with the delete operator, which calls the appropriate destructor.

```
delete prPtr1;
delete prPtr2;
```

Unlike a constructor, a destructor does not take arguments.

You can use the new operator for built-in types also. These types are often used to allocate an array. Square brackets indicate the size of an array; round brackets pass constructor arguments.

```
char * cPtr = new char[10]; // allocate 20 char's
int *iPtr = new int[20]; // allocate 20 integers.
```

With the introduction of new and delete, C++ programmers should never need to call the malloc() and free() functions, although as with all standard C library calls, they are still available for backward compatibility.

7.1.2 Inheritance

Your program may contain many classes, and many of them have several data members and member functions in common. You can remove the common parts from each class definition and declare it in a single class. Perhaps you have a system with several dials, including a dial to set a temperature alarm and a dial to set the motor speed. What the two dials have in common is the ability to read an analog-to-digital converter (ADC), which interprets the position of a potentiometer connected to the dial. The ADC provides a value in the range 0 to 255.

```
class Dial
{
private:
    char *adcLocation;
    int value;
public:
    void updateValue(void);
    int getRawCounts(void);
};
void Dial::updateValue(void)
{
    value = readADC(adcLocation);
    assert (value >= 0);
    assert (value <= 255);
}
int Dial::getRawCounts(void)
{
    return value;
}
```

Various dials interpret the data in different ways, converting it to the appropriate units. You would now create a SpeedDial and a TemperatureDial, each with slightly different behavior. You can reuse the Dial class by including it as a data member of the SpeedDial class:

```
class SpeedDial
{
public:
    Dial dialPart;
    int getSpeedRPM(void);
};
```

```
int SpeedDial::getSpeedRPM(void)
{
    int speedEntered;
    speedEntered = dialPart.getRawCounts() * RPM_CONSTANT;
    if (speedEntered > MAX_SPEED)
    {
        return MAX_SPEED;
    }
    else
    {
        return speedEntered;
    }
}
```

Now the `dialPart` can be accessed as a member of the `SpeedDial`. If you create an instance called `motorSpeed`, you can access the `dialPart` as follows.

```
SpeedDial motorSpeed;
motorSpeed.dialPart.updateValue();
```

This is straightforward when there is one level of containment, but it gets more cumbersome as the number of levels increases.

Chapter 5 shows how a pointer to the start of a structure can be cast to a pointer to the first member. This technique allows you to cast pointers to `SpeedDial` to a pointer to a `Dial` in order to put the `SpeedDial` on the same list as a `TemperatureDial`. The following array can hold pointers to either

```
Dial *dialList[10];
dialList[0] = (Dial*)&motorSpeed;
```

This is awkward and assumes the programmer will not place data members before the `dialPart` field. The casts are also unsafe. In C++, this is not guaranteed to work for classes with virtual functions, which I shall discuss shortly. C++ provides a more elegant way of handling this problem, called *inheritance*.

When using inheritance, the "is a" relationship between `SpeedDial` and `Dial` is reflected more closely. Once `SpeedDial` is declared to inherit from `Dial`, then `SpeedDial` contains all the data and member functions contained in `Dial`, without referencing the contained object explicitly. When the relationship is represented this way, the `Dial` class is the base class, and the `SpeedDial` is the derived class. You can declare this as follows.

```cpp
class SpeedDial : public Dial
{
public:
    int getSpeedRPM(void);
};
```

Relating the two classes in this way means that any access legal for a Dial is also legal for SpeedDial.

```cpp
Dial *dialList[10];
dialList[0] = &motorSpeed;
motorSpeed.updateValue();
```

A pointer to a Dial can also point at a SpeedDial with no cast. The member functions of the Dial, such as updateValue(), can be accessed directly through the SpeedDial. This is an example of polymorphism. Sometimes, it is a Dial; sometimes, it is a SpeedDial. It can morph, or change, from one to the other with no extra programming.

The implementation of getSpeedRPM() has also changed slightly. The dialPart does not have to be explicitly referenced. The call to getRawCounts() automatically accesses the base class.

```cpp
int SpeedDial::getSpeedRPM(void)
{
    int speedEntered;
    // the call to getRawCounts() is not preceded by an object name
    speedEntered = getRawCounts() * RPM_CONSTANT;
    if (speedEntered > MAX_SPEED)
    {
        return MAX_SPEED;
    }
    else
    {
        return speedEntered;
    }
}
```

The TemperatureDial can inherit from the Dial in the same way. It has its own functions that allow the caller to access setting information, getCentigrade() and getFahrenheit(), which allow the caller to read the temperature in either unit.

```
class TemperatureDial : public Dial
{
public:
    float getCentigrade(void);
    float getFahrenheit(void);
};
float TemperatureDial::getCentigrade(void)
{
    return getRawCounts() * CENTIGRADE_GAIN - CENTIGRADE_OFFSET;
}

float TemperatureDial::getFahrenheit(void)
{
    return (getCentigrade() - 32.0) / 1.8;
}
```

Figure 7.1 shows the relationship between the dial types. This inheritance diagram is only one level deep, but inheritance hierarchies several levels deep are not uncommon. Because a derived class can be thought of as containing both the base class and the derived class, both constructors are called, one to initialize each class. Similarly, when the object is destroyed, both destructors are called.

7.1.3 Virtual Functions

The real power of inheritance comes into play when you use *virtual functions*. A virtual function is a function defined in the base class and also defined in several derived

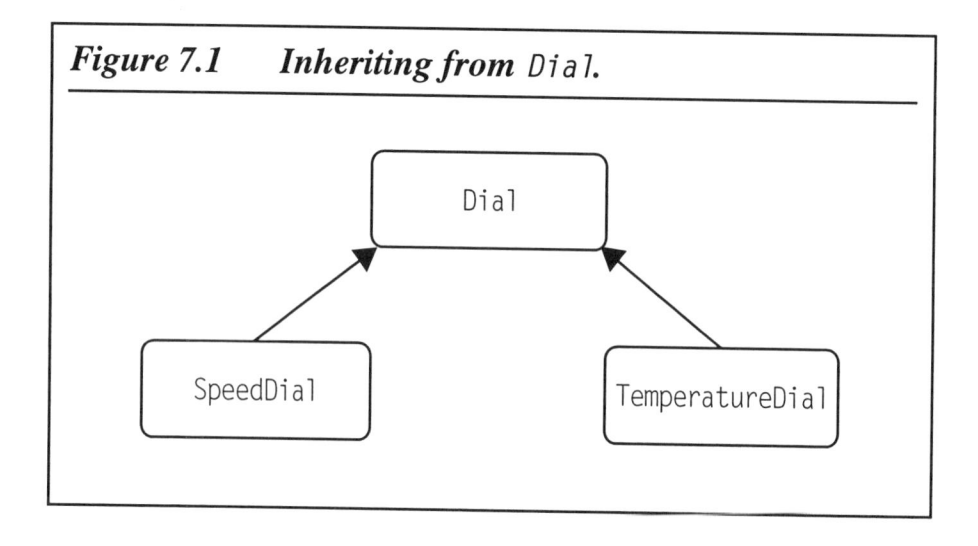

Figure 7.1 *Inheriting from* Dial.

classes. When the function is called via a pointer to the base class, the appropriate function is called, matching the actual derived type. For example, assume that you want each dial to perform an action whenever it reads a new value. You already have an updateValue() function, but it performs the same action for every dial. By rewriting this function for each class, you can perform a different action for each class' response to the same function call.

You must first declare the function virtual in the base class. To do this, the keyword virtual must appear at the start of the signature within the class definition.

```
class Dial
{
private:
    char *adcLocation;
    int value;
public:
    virtual void updateValue(void);
    int getRawCounts(void);
};
```

Now the derived classes also add this signature to their class definitions and write a new version of the function.

```
class SpeedDial : public Dial
{
public:
    int getSpeedRPM(void);
    virtual void updateValue(void);
};
void SpeedDial::updateValue(void)
{
    Dial::updateValue();
    setMotorSpeed(getSpeedRPM());
}
```

The first line of this new updateValue() function calls the parent updateValue() function. You still get all the functionality of the parent function, but you can add functionality specific to the derived class. This is not always desirable; in some cases, you may want to replace the base class functionality completely. But in this case, you want to add to it. Once the base class reads the ADC, the SpeedDial calls a function to update the motor speed. Obviously, this functionality is unique to the SpeedDial. The TemperatureDial may choose to add a different behavior, or it may choose not to override the virtual function, in which case the base class function is called whenever updateValue() is invoked on a TemperatureDial. Now you can loop through an array of dials, without caring about the actual type of each element.

```
SpeedDial motorSpeed;
TemperatureDial temperatureInput;
Dial *dialList[10];
dialList[0] = &motorSpeed;
dialList[1] = &temperatureInput;
/* initialize rest of the list in some way */
for (int i=0; i<10; i++)
{
    dialList[i]->updateValue();
}
```

The call to `updateValue()` can execute a different function each time around the loop, depending on the type of object pointed to in the array. This is the real power of virtual functions. They allow objects to look the same from the outside but behave differently in response to a function call. When the actual function called is not decided until run time, it is known as *late binding*. Normally, the exact destination of a function call in C is known at compile time, and thus is known as *early binding*.

7.1.4 Templates

Inheritance and virtual functions allow the behavior associated with a function name to vary according to the object type. Templates look at the other side of the coin. What if you want the functionality to stay exactly the same but to change the data type that is manipulated? This is often the case for code that sorts or manages lists. A simple algorithm can be applied to integers, but what happens when you want to sort floating-point numbers? It is not elegant to copy all the code, substituting `float` for `int` in each place it occurs. Templates provide a solution by allowing you to write a function with a type as a parameter. The template is invoked for each type required. I use the `max()` function as a simple example. The original C function is as follows.

```
int max (int a, int b)
{
    if (a > b)
    {
        return a;
    }
    else
    {
        return b;
    }
}
```

A `max()` function that takes two `float`s and returns a `float` applies exactly the same algorithm. When you want a function for `long int`, you repeat it again. In C++, you can define the > operator for a class, so you may want to define the `max()` function for a class. As the list gets larger, repeating the code and changing the type each time becomes less attractive. C++ allows you to rewrite the code shown earlier in a more general way.

```
template <class Type>
Type tmax(Type a, Type b)
{
    if (a > b)
    {
        return a;
    }
    else
    {
        return b;
    }
}
```

The first line tells the compiler that for the following code, the parameter `Type` is a data type — the class keyword is misleading because the data type may be a class, but it may also be one of the built-in types. Whenever `tmax()` is called, the compiler looks at the argument types and generates the appropriate function. The following two calls create two slightly different functions at compile time.

```
int a = max (5, 8);
float f = max (3.7, 9.1);
```

In most cases, you do not form a template of a function, but of a class. The following class holds an array of 10 pointers to objects of any type you wish to define.

```
template <class ElementType>
class SetOfTen {
private:
    ElementType *array[10];
public:
    ElementType *getItem(int index);
    void store(ElementType *elementPtr, int index);
};
```

Now, any time a class of 10 pointers to integers is required, it can be declared as follows.

```
SetOfTen<int> tenIntPtrs;
```

A set of 10 pointers to dials can be declared as follows.

```
SetOfTen<Dial> tenDialPtrs;
```

Templates offer code reuse that differs from inheritance. Inheritance encourages you to take an existing implementation and add something to it. Templates encourage you to write a general solution, which can represent a whole family of implementations, and then allow the user of the template to choose the implementations he requires. C++ allows the two techniques to be mixed freely.

7.1.5 Exceptions

It is not always reasonable to exit the program if a function fails. Some failures can be resolved at a higher level, and the operation can be attempted again. A piece of communications software may fail because the link is broken. The caller may decide to retry, or the caller may decide that the program must exit. The lower level communications routine may not know which action is appropriate for each call.

You can use an error code as the return value or as one of the function's parameters as one way of implementing the logic that tells the caller the result of the operation. This approach works, but it limits the information that can be passed when the failure occurs. The error-checking code is also invasive because almost every function call is surrounded by an `if` clause checking the error code. This leads to code that is difficult to read. There is also no guarantee that you will remember to check the returned error code.

C++ allows the execution to jump from within a function to a point in the caller's scope, called the `catch` clause, which contains code whose only purpose is to handle error conditions. A parameter can be passed from the point where the error was detected to the `catch` clause. When the error is detected, the action that causes the jump to the error-handling code is known as a *throw*. Look at the following simple example.

```
void increment(int *x)
{
    if (*x == INT_MAX)
    {
        throw "int overflow";
    }
    *x ++;
}
```

Although the value thrown in this case is a `char *`, any data type can be thrown. Often, a class is created to contain the data required to fully describe the problem, but a string suffices for this simple demonstration. Now that you have seen how the error is thrown, see how it is caught.

```
void foo(void)
{
    int a = 5;
    int b = INT_MAX;
    int c = 8;
    try
    {
        increment (&a);
        increment (&b);
        increment (&c);
    }
    catch (char *errString)
    {
        printf( "error = %s", errString);
    }
    a = b;
}
```

The first call simply increments a with no error. The second call uses a value already equal to INT_MAX and, therefore, causes the exception to be thrown. The catch clause takes a parameter a bit like the parameter list to a function. Inside the catch clause, the parameter can be examined. After the catch clause is executed, the flow of control continues with the assignment of b to a. However, the last line of the try clause is skipped. The catch clause must ensure that the state of the program is reasonable, even if the try clause did not complete. Indeed, one of the function calls within the try clause may have partially completed, making a suitable recovery even more difficult. For this reason, a function may have several try/catch blocks. Each catch clause need only worry about the calls in its own try block.

You can also have several catch blocks. The following example uses two empty classes to distinguish two different errors.

```
class TooSmall {};
class TooBig {};
void displaySpeed(int speed)
{
    if (speed < MIN_DISPLAY_VALUE)
    {
        throw TooSmall();
    }
    if (speed > MAX_DISPLAY_VALUE)
    {
        throw TooBig();
    }
    /* rest of code to display the speed */
}
```

The two classes that identify the errors are empty in this case. If the error required further information, it could be placed inside the class. The `catch` clause for these two errors functions a little like a `switch` statement, except that this clause searches for a `catch` clause with a type that matches, rather than a value that matches.

```
void foo(void)
{
    try
    {
        displaySpeed(speed);
    }
    catch (TooBig)
    {
        displaySpeed(MAX_DISPLAY_VALUE);
    }
    catch (TooSmall)
    {
        displaySpeed(MIN_DISPLAY_VALUE);
    }
    catch (...)
    {
        printf("Unknown exception occured");
    }
}
```

If the type of the exception thrown does not match either class named, the ellipsis (...) clause matches any other exceptions this function did not expect to receive. The ellipsis clause is, effectively, the `default` clause.

This type of error handling is used in several ways. You can unwind the stack when an error occurs and output as much information as possible to debug the problem. In this approach, the exception always leads to the program exiting. Another approach is to try to recover. In the example just shown, it may not be possible to display the current speed because the value is larger or smaller than the display's capacity. The calling function can display the maximum or minimum value instead and then continue execution. The value displayed is the closest possible to the actual speed. This may not be ideal behavior, but the system can continue to function. You do not want to implement this logic inside the `displaySpeed()` function because this recovery policy may be suitable in some cases, but not in others.

Regardless of which approach you choose, exceptions should be used only to handle cases that are actual errors and would not occur in the normal running of a program. The flow of execution of an exception is not intuitive, and it can be difficult to follow in an inspection. For this reason, do not use exceptions to handle expected failures. By an expected failure, I mean a failure that is part of a normal path of execution. Consider a program that regularly searches a list. Sometimes, a matching item is found; sometimes it is not. Both cases are normal and should be reflected in the return value. Using an exception to handle the not-found case leads to badly structured code. On the other hand, if two elements on the list match the searched-for item, then this may show that the list has been badly formed and some sort of recovery is necessary This is a valid case for an exception. In short, think of using exceptions when you might have used an `assert` macro.

Use exceptions to reflect only program errors, not conditions that may occur regularly in your program.

7.1.6 Other Features

C++ delivers many other minor features. Some make it a little more attractive than C. Default arguments are a nice addition that allow you to omit some arguments in a function call. Consider the following signature

```
void drawBox(Box *b, Color c = BLACK);
```

This defines a function that takes two arguments. However, if the second argument is not supplied, it defaults to BLACK, so the function can be called with one or two arguments.

```
drawBox (boxPtr, RED);
drawBox (boxPtr);
```

The first call draws a red box, the second call draws a black box. If the default color changes, the change can be made in one location.

```
void drawBox(Box *b, Color c = BLUE);
```

Now all the defaulted calls draw the box in blue.

Other features appeared because of the problems that cropped up when programmers used the capabilities new to C++. Run-Time Type Identification (RTTI) allows the programmer to establish the type of a class at run time. This was not needed in C because the type of structure never changed. In C++, you can have a pointer to a `Dial`, but you may not know whether that pointer points to a `SpeedDial`, a `TemperatureDial`, or another type of `Dial`. The following assignment succeeds if the `Dial` d is actually a `SpeedDial`. If not then `sd` is set to NULL.

```
Dial *d;
/* d gets a value in some way */
sd = dynamic_cast(<SpeedDial*>(d))
```

This slightly ugly syntax allows you to try to cast d to a SpeedDial. In C, a cast always succeeds, but it may cause a bug if the types are not compatible. In C++, the RTTI allows you to check if the cast should succeed. This is not necessarily a huge improvement over C, because many of the reasons for converting from one type to another did not exist before inheritance was introduced. It is more of a housekeeping feature than a feature that enhances your design.

C++ has many other features, but this overview should give you the flavor of what the language has to offer. If you want to learn more but are not prepared to commit to writing a major project in C++, then try writing some small utilities or test programs. You can experiment without excessive risk.

7.2 Choosing C++

Is C++ appropriate for embedded systems? Some critics claim that any extra overhead is unacceptable, though many of these same programmers made the transition from assembler to C and accepted that performance penalty. These critics also use an RTOS and accept its performance cost in exchange for more maintainable code. On the other side, some programmers claim that C++ leads to a cleaner design, which in turn leads to greater efficiency because optimizations are made at a higher level. Evidence for this is anecdotal at best, and many projects that use C++ hoping for a clean design, end up with worse spaghetti code than would be possible in C. Because it is such a contentious issue, I will show my colors now so that you can interpret what is to follow in that light. I believe that C++ is a great language with a lot to offer, but its sheer size means that for any given project, some features must be left unused. On projects below a certain size, the tools and the learning curve for team members who are not already familiar with the language is not justified. So if C++ is chosen, you must accept that the payback will not come until future projects reuse some of the code and some of the same people.

7.2.1 C++ for User Interfaces

User interfaces lend themselves to object-oriented programming because of the modularity of the components with which the user interacts. C++ also puts a lot of emphasis on designing the program's data, compared with designing the algorithm. In a user interface, the user can indirectly alter data within the program, so designing and grouping that data well can lead to a better overall design. By comparison, a feedback loop for process control may have a numerically complex algorithm that operates on relatively few items of data.

C++ also works well for user interfaces because it is a good match for event-handling algorithms. Because you can pass ownership of an event from one object to another, you can easily track which objects changed state due to an event and which did not. In some cases, an event is passed to several objects in turn. Each object can consume the event or return a status that says the event was not appropriate to this object, and the next object on the list can be tried. The search terminates when an object on the list successfully consumes the event.

The case for C++ is even stronger in GUIs. One reason is that a dialog's graphical controls may need to come into existence when the dialog appears and be removed from memory when the dialog is replaced by another. You cannot allow them to exist simultaneously because there may be many possible dialogs, and maintaining all of them in memory is not feasible. Such dynamic creation and deletion of objects is handled smoothly by the new/delete operators and the class constructors/destructors.

7.2.2 Programming in the Large

The biggest consideration when choosing to go the C++ route is the project size. The code should be big enough to benefit from the improved encapsulation. The budget should be generous enough to hire or train a team so that it has adequate C++ expertise. The target device memory should be big enough to carry the extra weight of C++ resource-hungry features. You can write C++ code that is as efficient in size and speed as C. But if you choose this route, you lose the advantages of C++. C++ allows you to trade efficiency for maintainability — the same trade-off you face in the transition from assembler to C.

As the number of programmers grows, it is more important to be able to find the home of a piece of functionality. Classes provide the encapsulation that allows one piece of functionality to be updated, with limited consequences for the rest of the program. Like any tool that facilitates better structure, C++ gives you places to put things. The places are classes and the things are data, code, type definitions, and constants.

The following anecdote illustrates the problem that C++ tries to solve for large programs. During a discussion on C++ with a colleague, I was using a C program he had written to illustrate how it could have been designed in C++. The program allowed the user to display individual characters of a bitmapped font, magnified to make individual pixels visible. The user could edit those characters to form a new character. At the bottom of the screen was a font preview area — a single line to allow some characters of the font to be displayed at their normal size. I explained that C++ would allow a separate class for each area of the screen, and that class would contain all the code and data associated with area. Because the program was incomplete, the user had to press Return at the end of a sequence of typed characters before the characters would appear in the font preview area. It would be more desirable to have each character appear as it was typed rather than having to type blind until the Return key was pressed. I used this as an example, saying that when it came time to rectify this shortcoming, the place in the code to do the updates would be in the class associated with that area of the screen. This would be the appropriate place to put the new code, and the change would be localized. His response was, "I can fix that problem in a single line of code. Watch this." The single-line change did the trick, but the line was in the main() function! In a small program, doing an update to main() is acceptable. But it becomes intimidating to change a line of code in the main() function in a large program, as a change could have unexpected implications anywhere in the program. The number of code paths and possibilities to be considered when deciding whether the change is acceptable grows exponentially. This may be an extreme case, but it illustrates that C++ gives the programmer a place to put things. Where these things — data code, type definitions, constants, and so on — are placed matters little in a small program, because a brute-force search can always uncover them. But as the program grows, navigation becomes difficult.

Name collisions are another problem with very large programs. In C, the function addToList() in one module can collide with addToList() in another module, which is dealing with a completely different list. Classes alleviate this problem because the member names are meaningful only when used with an object of that class. C++ has another feature, called *namespaces*, which allows a set of types, functions, and data to be prefixed with a particular name in order to further protect a library or module from name collisions.

Is your embedded system large? You should not always measure the size by the number of lines of code. When choosing the language, the number of programmers is more important. This is what decides how many paths of communication there will be between programmers, and it is what decides how many people will have to be familiar with a function. Although it is true that if there are 10 programmers, not all must be familiar with all of the code, there will still be sections that four or five of them must know. Isolating pieces of the design in classes minimizes the time required to understand the workings of a previously unseen function. You cannot say that C++ is worthwhile after 40,000 lines of code or that it works better after the number of programmers is more than eight. The decision lies on a gradual scale that must include the other contributing factors discussed in this chapter.

7.2.3 The Performance Question: Does C++ Really Lead to Bigger, Slower Executables?

Like many of the tough questions in computer science, the answer is "it depends." C++ was designed so that if you do not use a feature, you should not pay for it. Thus, a class with no virtual functions occupies the same memory as a structure in C containing the same data elements. This approach should mean that if a C program is compiled with a C++ compiler, the performance and resources used should be identical. Many compiler vendors digress from this philosophy in the areas of exceptions and multiple inheritance, but these changes in performance are unlikely to affect your application. In any case, if you do not use these features, you will probably disable them by passing a switch to the compiler and thus removing all overhead.

A larger difference in performance stems from the type of design that C++ encourages: object-oriented design. Encapsulation leads to many and smaller functions. Dynamic creation of objects with short lives forces the heap to work hard. General solutions that encourage code reuse can lead to bigger classes, which lead to bigger programs in cases in which only a small amount of the classes' functionality is required. So programs that are designed according to the C++ philosophy suffer a performance and resource overhead, not the language itself. Same thing, you might say. It is not the same thing, because you have control over your program; you do not have control over the language itself. In the next few sections, I look at some coding practices and discuss some solutions.

7.2.3.1 In-Line Functions

Using encapsulation forces you to create functions to access data members, so a C++ program tends to have many small functions, leading to slower execution and greater stack-depth requirements. One newsgroup anecdote describes a case in which a program took 30 nested function calls to toggle a single bit.

To counter this, in-line functions allow optimization of many small routines. Normally, each function exists at a single memory location, and at each call, the compiler inserts a jump to that location, having pushed the arguments onto the stack. An in-line function is one in which the compiler expands the function in place, much like a macro in C. Typically, the in-line function is defined in the class definition.

```
class Setting : public ArrowCallback
{
private:
    int value_;
public:
    /* other functions declared here */
    int getValue(void)
    {
        return value_;
    }
};
```

By including the code in the class definition (which is typically included in a header file), the `getValue()` function can be expanded in place to a simple access to `value_`. In this case, the code will be smaller and faster, because the generated code to access the member is smaller than the code required to put parameters on the stack (the object pointer goes on the stack even if there are no other parameters defined) and to jump to the subroutine. If the contents of the function are more complex, expanding it many times is still faster than a call. But the code will grow in size. Beyond a certain size, most compilers refuse to in-line the function.

180 — Front Panel: Designing Software for Embedded User Interfaces

Note that the in-line function is superior to a macro. A macro cannot access private members of the class. Also, a problem with macros is that the arguments are expanded each time they occur in the definition. Consider the following macro.

```
#define MAX(a,b) ((a>b) ? (a) : (b))
```

The arguments appear twice, so the following call increments x twice.

```
z = MAX(x++, y);
```

In-line functions do not suffer from such side effects.

Although in-line functions are an improvement over macros, they generally replace functions that would not exist in the equivalent C program, because the C program would simply access the structure directly. In-lining goes some way toward compensating for the proliferation of functions in C++ code, but it has restrictions. The compiler may not always be able to in-line a function if it is virtual. More significantly, a change to an in-line function requires recompilation of all modules that include it. For this reason, you should program with few or no in-line functions initially and convert some of the out-of-line functions to in-line functions when profiling and optimizing the program.

7.2.3.2 Code Reuse Effects on Efficiency

C++ promotes code reuse and thus encourages bigger classes that are more general in purpose. For example, if you write a linked list that can be used for any linked list that you may ever need, that implementation will be bigger and slower than an implementation for one special case. The payback happens only if the general-purpose linked list is used frequently and the total amount of list-managing code is reduced. Total

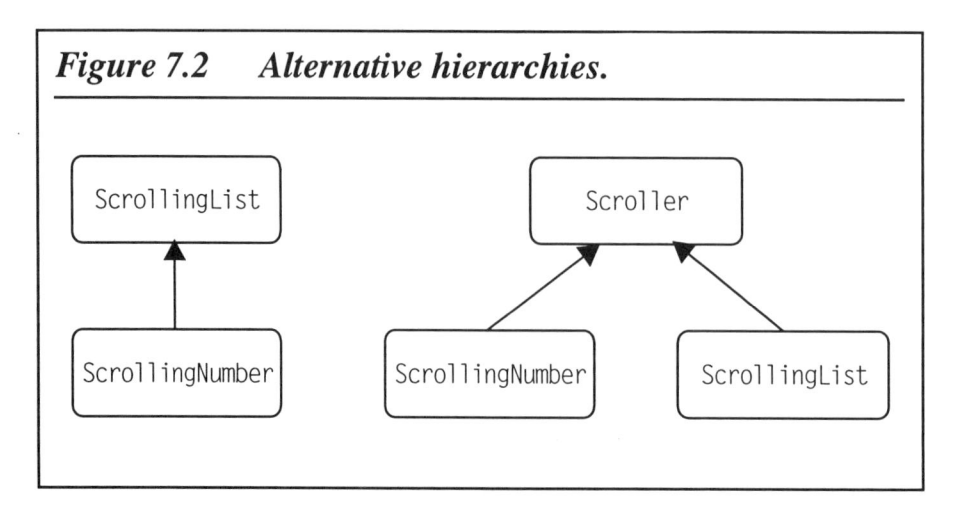

Figure 7.2   Alternative hierarchies.

code size shrinks because you no longer have to write new code for each new list. Sometimes, embedded systems programs are simply not large enough to take advantage of this economy of scale.

Another reuse danger is that inheriting from an object involves adding the size of the base class to the size of the derived class. If most of the functionality of the base class is replaced with virtual functions, the data in the base class may never be accessed, but it still takes up space.

Sometimes, you can adjust the hierarchy. Consider developing a `ScrollingList` that iterates through a set of names on a text display. It has methods `scrollUp()` and `scrollDown()`, which update the display. Later you decide you want a method that allows the user to scroll through numbers. You call this `ScrollingNumber`. You inherit from `ScrollingList` because you still want to be able to call the `scrollUp()` and `scrollDown()` functions. So the inheritance diagram looks like the left-hand side of Figure 7.2. Unfortunately, all the names stored in the `ScrollingList` still have space allocated for them in the `ScrollingNumber`. To avoid this, create a base class from which both can be derived, as shown on the right-hand side of Figure 7.2. The base class may contain little or no data, but it defines the functions `scrollUp()` and `scrollDown()` so that they can be redefined in both derived classes.

Reorganizing classes in this way avoids the most blatant waste of memory, but there are cases in which you inherit from a class and only some of its data is required. In such cases, you must decide between code reuse and memory consumption.

7.2.3.3 Memory Management

One of the strongest criticisms of C++ in the embedded systems community is that it is a memory hog and, worse, that it is prone to memory leaks. These are completely unacceptable in an embedded system that may run for months without a reset. The truth is that while object-oriented programming encourages dynamic object creation and deletion, C++ gives you more options for memory management than C, compensating for this danger.

Even if you decide in a C program that you will not use the heap and allocate everything statically, you can still do so in a C++ program. Instead of allocating the structures statically, you allocate the objects statically. In this case, the constructors of the objects must be designed so that they do not depend on initializing the system before they run. If a class is declared statically, its constructor runs before `main()`, and you cannot guarantee the order of execution of constructors for objects declared in different modules.

7.2.3.3.1 Operator new

C++ replaces `malloc()` with the operator `new`. You can override this operator to customize the allocation mechanism. It can simply abort, ensuring that if anyone attempts to use the heap in the program, it will be detected immediately. This is useful if you wish to allow only statically allocated data.

The operator new can also be overloaded on a per-class basis (see Listing 7.1), allowing you to write a memory allocation scheme to suit a particular class. This is useful if you decide to use the new and delete operators freely, but you know that there is an upper limit to the number of objects of a certain type that can exist at any time. The following class achieves this by providing new and delete operators that use memory from a statically declared array.

As well as new and delete operators, C++ separates operators for arrays, known as new[] and delete[]. They are invoked when the program allocates and frees arrays on the heap.

```
Foo *fPtr = new Foo[4];
delete[] fPtr;
```

This code allocates an array of four Foo objects and then deletes them. Using the following nonarray delete operator is an error, because only the destructor of the first of the four Foo objects is called.

```
delete    fPtr;
```

Memory allocated as an array should be freed as an array, and memory allocated as a single object should be freed as a single object. This is something you have to be careful of in C++, or you might get some very mysterious bugs.

When implementing operators new and delete on a per-class basis, you never want to use the array operators, so you can disable them by declaring them private. This leads to a compile-time error if code outside the class tries to allocate or delete an array, and it causes a link-time error if called from inside the class because the function is never actually defined.

Listing 7.1 The Foo class that implements its own heap management.

```
#define POOL_SIZE 10
class Foo
{
int i_;
    /* some other data */
    /*
    In C++, the argument name can be dropped if it is not used,
    so size_t is the type passed in but never used.
    */
    void *operator new[](size_t);
    void operator delete[](void *);
public:
    void *operator new(size_t size);
    void operator delete(void *fooPtr);
};
```

Listing 7.1 (continued)

```
unsigned char GS_fooHeap[POOL_SIZE][sizeof(Foo)];
/*
Mapping a bit to each block would be more efficient than using
a Boolean that is a byte, but this will suffice for the example.
*/
Boolean GS_blockUsed[POOL_SIZE] = {FALSE};
void *Foo::operator new(size_t size)
{
    for (int i=0; i<POOL_SIZE; i++)
    {
        if(GS_blockUsed[i] == FALSE)
        {
            GS_blockUsed[i] = TRUE;
            return GS_fooHeap[i];
        }
    }
    /*
    If the program reaches this point, we have tried
    to allocate more Foo objects than POOL_SIZE.
    */
    assert(FALSE);
}
/*
By the time this function is called, the destructor has already
been run, if there is one.
*/
void Foo::operator delete(void *fooPtr)
{
    /*
    Deleting a null pointer should have no effect.
    */
    if (fooPtr == NULL)
    {
        return;
    }
    /*
    Figure out where this block is in the array.
    The void *'s are cast to Foo *'s so that the pointer
    arithmetic is done on elements of sizeof(Foo).
    */
    int blockIndex = (Foo *)fooPtr - (Foo *)GS_fooHeap[0];
    /*
    Ensure that the index is legal. A careless programmer may
    have tried to delete an object that was allocated on the stack.
    */
    assert((blockIndex >=0) && (blockIndex < POOL_SIZE));
    GS_blockUsed[blockIndex] = FALSE;
}
```

If classes maintain separate pools of memory, they do not prove as efficient as a single heap, but the problem of fragmentation is avoided. It is also easier to either verify the absence of leaks or identify a leak's exact location. You can also disable the global `new` and `delete` operators by defining them to always exit the program. This prevents the program from doing any heap allocation other than that managed by the classes. Allocating and deleting single objects can now be performed as before, but array allocations are prevented.

```
Foo *f1Ptr = new Foo;
Foo *f2Ptr = new Foo[3]; // syntax error - trying to access private
                                       member.
delete f1Ptr;
delete[] f1Ptr; //syntax error again
```

If the operator `delete[]` had not been disabled, the last line would not be a syntax error. But it would have been a serious logical error. The space allocated by the custom `new` operator would be returned to the global heap, causing heap corruption.

Disabling allocation mechanisms may seem a bit draconian — I mean, you do trust the programmers. Don't you? When you consider large programs with new staff coming on board regularly, this kind of protection goes a long way toward preventing errors. It is also appropriate in this case because the rules for the subtle differences between an array `delete` and a single object `delete` are not intuitive and cannot be remembered easily.

The power of mechanisms like the one just shown allows you to tame C++ memory management. If you gave memory management serious consideration in C, you will do so in C++. If your demands on the heap are restricted, you may be better off with C++ than with C.

7.2.3.3.2 Heaps and Fragmentation

If you allow full use of the heap in C or C++, fragmentation is a problem, even if you are cautious and allow no memory leaks. Typical use of the C++ `new` operator tends toward many small allocations. A heap area scattered with many small allocations fails if a request is made for a single large contiguous block that is no longer available. As an aside, successful allocations in fragmented heaps have nondeterministic execution times, because the list of free blocks must be searched until one is found that is large enough. This may matter to the hard real-time parts of your program, but it will probably not be an issue for the user interface software.

One reason for small allocations is that object constructors often allocate other objects that are then pointed at by data members. In some cases, making one of the objects contain a second object rather than a pointer to it leads to one large allocation instead of two small ones. However, the number of cases in which this applies is small.

To allow full heap use with no fragmentation, you can overload the global new operator to provide many pools of allocation blocks. Rather than having one pool per class, you can have a range of pools. The blocks in each pool are of a different size. When a memory request is made, a block of the nearest larger size is handed out. Figure 7.3 shows three pools of different-size blocks. Some tuning may be required to obtain appropriate pool and block sizes for your application. This scheme prevents fragmentation problems in long-running systems but does nothing to prevent memory leaks from occurring if you allocate memory and never free it.

When you do have memory leaks to track, class-level new and delete are better than size-based pools, because knowing exactly the type of object that has been leaked is a good first step to finding the leak itself. You can define a global allocation scheme to allocate memory according to pools and to use class-based new and delete operators to count the number of objects of a particular kind that have been created and deleted. The class-based operator new can record auditing information and then explicitly call the global operator new to allocate the space.

7.2.3.3.3 Tasking

The per-class operators new and delete that I defined do not provide thread-safe code. If different tasks can allocate objects from the same pool, you must perform some locking to protect the shared data structures from simultaneous access. You can do so by locking an operating system resource at the start of the new and delete operations and releasing it at the end.

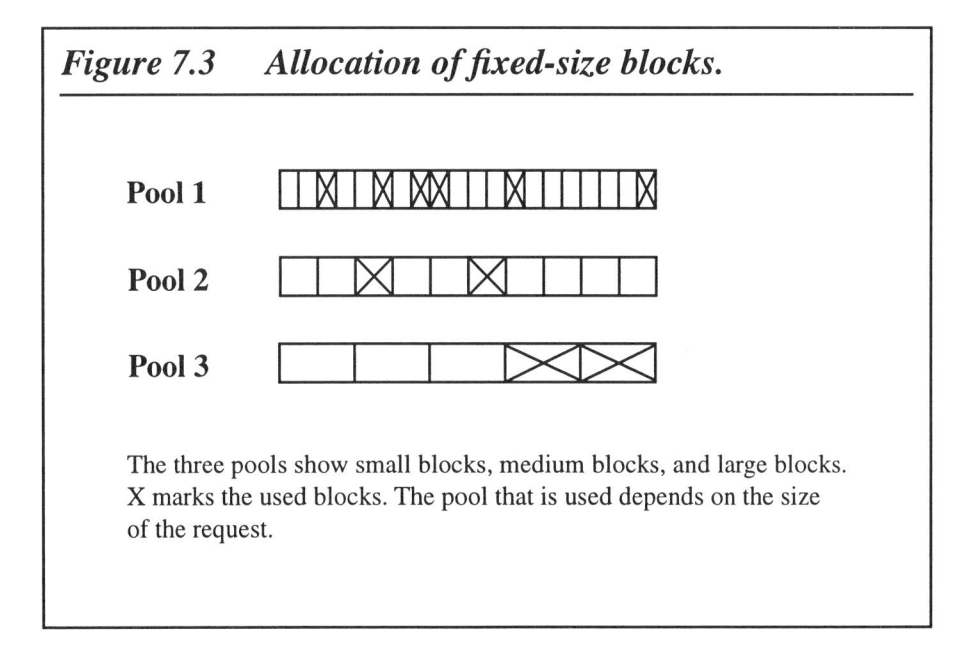

Figure 7.3   Allocation of fixed-size blocks.

Pool 1

Pool 2

Pool 3

The three pools show small blocks, medium blocks, and large blocks. X marks the used blocks. The pool that is used depends on the size of the request.

Another issue is that the tasking system may dynamically create and delete tasks. If a task is halted while running, it may have memory allocated. You may need to keep objects in a per-task pool so that they can be tidied up when the task is halted. This issue is addressed in more detail by Lethaby and Black (1993).

7.2.4 Code Reuse and Complexity

C++ strives to reuse code at several levels. At the simplest level, a piece of code should be used many times within the same program. Another level of reuse occurs when several classes are used in a set of programs within a team or company. C++ encourages code reuse at the level at which the writer tries to construct a class that can be used in a general way. The original intention was that class libraries could be bought and sold or that libraries for a specific area could be reused throughout an organization. This led to the requirement that classes be usable when the source code of the library is not available. If not for this requirement, certain features in the language would not be necessary. Those features can lead to increased coding complexity if used widely. I will examine some of these features and then draw conclusions for embedded systems.

7.2.4.1 Multiple Inheritance

Multiple inheritance was introduced partly to allow application objects to have inheritance relationships with classes from different libraries. Some class libraries can be used properly only via inheritance. If library A demands that your application class inherit from BaseA and library B demands that your application class inherit from BaseB, then your application class must inherit multiply. If all source code were available and editable, a cleaner solution might be found — for example, BaseA and BaseB could be merged, or one of them could inherit from the other.

7.2.4.2 Templates

A template is a feature that promotes general-purpose reusable code. The extra complexity of managing templates pays off only if the template is used for many different types. For example, if you write a simple list template but use it only for two types, then the extra effort of writing the list code as a template was probably not worthwhile. You are better off writing two different sets of code and having the option of optimizing each list for its own purpose. On the other hand, if you have 50 types of lists, the extra work of building a template pays off handsomely. Ideally, you simply buy the list template from a third party. I discuss this situation shortly.

7.2.4.3 Exceptions

Exceptions are another library-oriented mechanism. They are required because a library cannot decide the appropriate reaction to a failure in one of its functions, which might be due to the data passed in from the application. If you try to remove an element from a list and the list is already empty, then the application must decide whether it is an error or not. If all the source code is available, you can decide it is not acceptable and abort the program from the list code, probably by using an `assert` macro. Alternatively, you can decide that the exception is acceptable and have a non-exception return, maybe returning `NULL`, to indicate that the item is not present. Exceptions give control back to the caller when there is a failure in a library routine. Handing back this control is not necessary if you can decide what to do at the point at which the failure occurred.

In embedded systems, exceptions may seem like a good match for hardware failures. When you look closer, this is not always the case. When hardware fails, you decide either to shut down the system or to soldier on, regardless. In the first case, there is no point in passing control to the caller of the function, because you know you will jump to a tidy-up routine to shut-down or reset the system. In the second case, you must write the software so that all features perform in a well-specified way when the hardware is not available. In such a case, the routine accessing the hardware must indicate failure — not just when the failure occurs, but also during future accesses. This leads to a dangerous situation in which a function is called on a regular basis and throws an exception to indicate failed hardware on each iteration. This is not desirable because the path through the exception handler is slower than conventional returns and branches. More importantly, it is very difficult to write code that eliminates resource leaks, because anything allocated before the exception must be freed in the `catch` block. The `catch` block does not always know exactly where it was called from — hence the difficulty in knowing which resources were allocated. I am not saying it cannot be done, but if you are going to make that effort, you should first be sure that the effort has some payback.

You should note that two of the later and lesser-known features in C++ were added to resolve memory leak problems that arise when using exceptions. The first is the `auto_ptr`, which lets you track objects created during a constructor so that they are destroyed if an exception is thrown later in the constructor. The second is the placement `delete` operator, which can free memory that is being allocated by an overloaded `new` operator when an exception is thrown. If code using exceptions were straightforward, these issues would have been recognized as necessary when the exceptions feature was first introduced. If language designers have trouble predicting what sort of problems exceptions will cause, there is little hope for the rest of us. Worrying about such extra features significantly detracts from the simplicity and, therefore, the usefulness of exceptions. Cargill (1994) gives further examples of the difficulty of writing exception-safe code.

If your device's specification dictates that you must tolerate certain hardware failures, then exceptions will not make your job easier. You are better off treating that requirement the same way you treat the requirement that the system should handle invalid operator key presses: write conventional paths to follow for all the possible failures. Your code will be clearer, more testable, and more maintainable. Exceptions have a role in detecting hardware failures when you want to write a library that uses the hardware but you do not know the failure-handling policy of the applications using the library. However, such situations are rare in embedded systems.

7.2.4.4 Taming the Complexity

I have said that the multiple inheritance, templates, and exceptions features have their real place in commercial third-party libraries that sell general solutions in some problem domain. This market has not reached its full potential in the desktop programming arena, though there are notable successes, such as the standard template library (Musser and Saini, 1996) and a number of graphical libraries.

With little success on the desktop, the embedded market will see even less activity in this area. Embedded systems' needs are often very specific. Even if the classes are thread safe, they may have to integrate with a variety of RTOSs to be reusable. Possibly the most important factor is that small embedded systems cannot afford to link in a large third-party library if they will use only a fraction of the functionality provided. The philosophy of C++ is that if you do not use it, you should not have to pay for it. But the same philosophy does not extend to all libraries. In theory, the linker allows you to link only the functions you actually use. In practice, many libraries have so many interdependencies that using some of the libraries' simpler functions produces a large overhead.

The embedded world will not benefit widely from reusable general-purpose libraries of classes, so embedded programmers should avoid paying the price for the features that are provided for that reason. Part of this price is performance and memory consumption. Plauger (1996) makes a case for restricting C++ for performance reasons. I believe that program complexity is a far stronger reason to restrict the use of features invented to serve criteria that do not apply to most embedded systems. Complexity must be taken very seriously in embedded systems. Complexity is not just a case of training the programmers to handle code that is structurally more difficult. Economies of scale dictate that embedded programming tools are always more primitive and buggy than those for desktop systems. Embedded systems have multitasking and real-time constraints that contribute to complexity, and require special tools for those problems. Adding further to the complexity only increases the risk of hitting a show-stopping bug that can cost companies in dollars — and programmers in lost sleep. Using C++ wisely reduces that complexity and risk, but exercising all the C++ features, without justification, leads to a system that is no longer understood or trusted.

7.3 Using C++

If you choose to use C++, you will need to adapt your implementation techniques in certain areas. In the following sections, I review the commonly used techniques whose implementation changes when you move from C to C++.

7.3.1 Implementing a Focus, or Callbacks, in C++

In Chapter 3, I discussed the use of a focus to track the object, or structure, that is affected by actions from a piece of the user interface — for example, trackball or shared arrow keys. In C, the focus can be implemented with a pointer to a piece of data, and callbacks are implemented with pointers to functions.

In C++, the two techniques merge because the same implementation suits both situations. It is known as a *callback* because that is the more general case. It is appropriate to point at an object rather than at a function. The object contains both the code and the data required to handle the event. Similar to the pointer-to-function technique, a signature must be chosen for the function to be invoked when the event occurs. If you have two arrow keys, the functions is as follows.

```
void arrowUp(void);
void arrowDown(void);
```

Although there are no arguments, some data may be passed from a motion device, such as a trackball. These two functions can be put in a class whose only purpose is to define the interface to an object that can be the target of a callback.

```
class ArrowCallback
{
public:
    virtual void upArrow(void) = 0;
    virtual void downArrow(void) = 0;
};
```

The = 0 at the end of the function declaration means that the function is *pure virtual*. This has two implications: first, objects of type ArrowCallback can never be created, although objects of types that inherit from ArrowCallback can be; second, any class that inherits from ArrowCallback must define the two functions in order to allow instances of the class to be constructed. The following is a simple derived class capable of storing and altering a setting.

```
class Setting : public ArrowCallback
{
private:
    int value_;
public:
    Setting();
    virtual void upArrow(void);
    virtual void downArrow(void);
    int getValue(void);
};
    Setting::Setting(void)
{
    value_ = 0;
}
void Setting::upArrow(void)
{
    value_ ++;
    if (value_ > MAX_VALUE)
    {
        value_ = MAX_VALUE;
    }
}
void Setting::downArrow(void)
{
    value_ --;
    if (value_ < 0)
    {
        value_ = 0;
    }
}
int Setting::getValue(void)
{
    return value_;
}
```

The upArrow() and downArrow() functions are given some meaning for the setting class. Other classes can interpret the arrow keys in different ways. If the arrow keys are used to navigate a menu, the menu class might inherit from ArrowCallback, but the upArrow() and downArrow() functions cause different text to appear on the menu's display.

Now you can define a class that manages the events for the arrow keys. The class knows who should receive the arrow events.

```
class ArrowKeys
{
private:
    ArrowCallback *target_;
public:
    void keyEvent(char key);
    void setCallback(ArrowCallback *target);
};
void ArrowKeys::setCallback(ArrowCallback *target)
{
    target_ = target;
}
void ArrowKeys::keyEvent(char key)
{
    if (target_ != NULL)
    {
        switch (key)
        {
        case UP_KEY:
            target_->upArrow();
            break;
        case DOWN_KEY:
            target_->downArrow();
            break;
        default:
            /*
            This key should not have been passed to this object.
            */
            assert(FALSE);
        }
    }
}
```

You now have a mechanism for pointing the activity on the arrow keys at any object inheriting from ArrowCallback and defining the upArrow() and downArrow() functions. You may have objects that receive callbacks from many sources, which may lead to multiple inheritance. Also, Setting could have already inherited from some other class to get to another part of its functionality.

ArrowCallback itself is a class with no functionality. In this case, inheritance is used to inherit an interface only. No data or implementation is inherited. If you must use multiple inheritance, multiply inheriting implementations cause far more problems than do multiply inheriting interfaces. Programs that use multiple inheritance widely are more complex than programs that do not, and genuine design reasons to use it are rare. The problems stem from ambiguities in casting, as well as ambiguities when base classes have members of the same name. The details of these problems are beyond the scope of this text, but they are discussed in detail in Eckel (1995). For these reasons, implementing callbacks should be one of the very few cases in which multiple inheritance is considered in your design, and even then it should be restricted.

7.3.1.1 C Callbacks from C++

If you program with a third-party graphics library, the library might force a different callback mechanism on you. If the library is C based, then it almost certainly forces you to use a pointer to a function. This function cannot be a member function of an object, because a member function would not match the signature dictated by the library. However, it could be a nonmember function within your C++ program, or it could be a `static` member function. `static` member functions associated with a class are not called on in any specific instance of the class. They have no `this` pointer to indicate the object with which they are associated. The `this` pointer is generally passed in secret to member functions to indicate the object that is acted upon, but not in the case of `static` member functions.

You may eventually want to get to an individual object. Most C libraries allow setting up the callback with a function pointer and a `void *` for miscellaneous data. The object pointer is stored in the `void *` and cast back to an object pointer in the static member function.

```
class CallbackFromC
{
public:
    static void callbackFn(void *miscellaneous);
    void action(void);
};
void CallbackFromC::callbackFn(void *miscellaneous)
{
    ((CallbackFromC *)miscellaneous)->action();
}
void CallbackFromC::action(void)
{
    /*
    This is where the real work in response to
    the Callback is performed.
    */
}
```

The `CallbackFromC` class has one static member function, which is called from the C library. The `action()` function is then called to do the real work. If you have to manage many callbacks, the `action()` function can be made virtual and any class that must handle a C callback can inherit from `CallbackFromC`.

`setCallbackInCLibrary()` is a function supplied by the C library to set up a callback. It takes arguments for the function pointer and the miscellaneous data. The C library did not intend this data to be used for an object pointer, but it is usually supplied to allow for application-specific needs. You can now set up a callback with the following code.

```
CallbackFromC targetObject;
setCallbackInCLibarry(CallbackFromC::callbackFn,
                     (void *)&targetObject);
```

The function name passes the address of the function, but the full function name includes the class name and the `::` separator, giving you `CallbackFromC::callbackFn`. A similar technique applies to interrupt routines, which are also called with no associated object.

7.3.2 Hardware Issues

Because you will probably be using C++ on custom hardware, you should know about C++ and that environment.

7.3.2.1 Start-up

One thing you may have to do is provide a way of calling the constructors of all statically declared classes. In C, statically declared structures simply set aside space. But in C++, constructors must run to initialize the statically declared objects. C++ demands that those constructors run before `main()` begins. If you can control your system's reset vectors, you should not jump straight to `main()` because you will miss the opportunity to construct those objects. Your compiler vendor should supply a function that will run each constructor. This function should be called before `main()`. A couple of `printf`s in the constructors may help establish if they are being called, but beware that `printf()` may not work until some of the standard I/O initialization has been performed, so stepping with a debugger may be a better option.

7.3.2.2 Encapsulation

At this stage, you should not be surprised that well-structured code hides the details of how hardware operates behind the public interface to a class. If the memory-mapped addresses are private, you know that no code outside the class accesses that device. If disabling interrupts or the locking of a resource is required, do the locking inside the member functions — you do not have to depend on the caller remembering to do some setup before accessing your class. The emphasis should be on doing extra work in the class to make life easier for the programmer using this class.

7.3.2.3 Classes in ROM

You should store strings in ROM to avoid wasting possibly scarce RAM resources. For example most compilers allow the following string to be placed in ROM.

```
static const char romString[] = "Hello World!";
```

Similarly, many structures with fixed-data, or lookup, tables should be placed in ROM.

As with structures, declaring an object to be `const` allows it to be stored in ROM, with a couple of provisos. The compiler generally raises a syntax error if any part of your program tries to alter the data in such an object. To take advantage of this feature, you should understand the ways in which C++ uses `const`. `const` means logically constant. This means that a pointer can be declared so that the data cannot be changed via this pointer, although it can change via other pointers which were declared without the `const` qualifier. The syntax for pointer to `const` follows.

```
const Foo *fooPtr;
```

Although `fooPtr` can point at many `Foo` objects, it cannot be used to change any of them. The following alternative means that the pointer `fooPtr` must be initialized to point at a single object and can never be changed to point at another.

```
Foo *const fooPtr;
```

This form is not relevant to placing classes in ROM and will not be discussed further.

If you have a pointer to a class and you are not allowed to change it, then how do you decide which member functions change the object and which do not? C++ allows you to place the keyword `const` in a member function prototype. This tells the compiler to forbid changes to the object within this function.

```
class sometimesConst {
public:
    void makeNoChange(void) const;
    void makeSomeChange(void);
};
```

This class can be pointed to by a pointer to `const` or a pointer to non-`const`. The pointer to non-`const` can be used to call either member function. The pointer to `const` allows only calls to the `makeNoChange()` member function that has the `const` qualifier.

This is a useful mechanism if you want to give read-only permission to a part of your program. By passing a pointer to `const`, you say, "Here is the means to access an object, but you cannot change it." The part of the program with full access to the object holds a normal pointer and can call const and non-const functions. The pointer to const, or the read-only pointer, allows access only to the const member functions.

Designing all your RAM-based classes to allow const and non-const qualified pointers to work consistently is a nontrivial exercise (Saks, 1996), but this section points out how your const objects can live in ROM. These objects must be declared const as in the following.

```
const Foo f;
```

Note that you declare the class, not the pointer to it. Another restriction is that the class, or its base classes, must have no user-defined constructors or destructors. By using casting, you can still write a program that attempts to alter data in this object. This leads to undefined behavior if the data is actually in ROM, so you must respect the const keyword. Even when these conditions have been met, you have no guarantee that your compiler will place the data in ROM (though most will). If necessary, you can take matters into your own hands with the placement new operator.

7.3.2.4 Placement new Operator

This section describes a fairly exotic use of the new operator. This use is not required in most programs, but it is occasionally applicable in embedded systems in which you tend to care about exact locations of data within memory. You may try to optimize use of on-chip RAM, because you want certain data to live in ROM, or because you have hardware located at a specific location in memory that you want to access via data members.

You can call new with a single argument after the keyword new, which is a pointer.

```
Foo *fooPtr = new (location) Foo(a,b);
```

a and b are arguments to the constructor of Foo; location is the argument to the new operator. Normally, operator new takes a single size_t argument. But arguments inserted after the new keyword also pass to the operator new function not to the constructor. The following version of operator new provides the placement functionality you desire.

```
void *operator new(size_t, void *p) throw ()
{
    return p;
}
```

This is part of the standard C++ library. Your compiler should supply it, but some compilers do not. In any case, it is trivial to include this function in your own code. By returning the pointer passed in, this function places the object at location p. Now the following call to new places the Foo object at address location.

```
Foo *fooPtr = new (location) Foo(a,b);
```

This address must be a pointer and may have been cast from a `long int` if the exact address is known

```
Foo *fooPtr = new ((void *) 0x1234) Foo(a,b);
```

Consider a simple text display of 20 characters which you can access via two consecutive bytes: a control byte and a data byte. The control byte is used to reset or erase the display. It is also used to set the cursor to an arbitrary position by setting the most significant bit of the byte and passing in the position in the other seven bits. The following class gives you a clean interface to the memory-mapped hardware.

```
class TextDisplay
{
private:
    typedef enum {RESET = 0x17, ERASE = 0x18, SET_CURSOR = 0x80 };
    volatile unsigned char control_;
    volatile unsigned char data_;
public:
    void reset(void);
    void erase(void);
    void setCursor(int position);
    void displayString(char *string);
};
void TextDisplay::reset(void)
{
    control_ = RESET;
}
void TextDisplay::erase(void)
{
    control_ = ERASE;
}
void TextDisplay::setCursor(int position)
{
    assert (position < 20);
    control_ = SET_CURSOR | (char) position;
}
void TextDisplay::displayString(char * string)
{
    /*
    Iterate through the string outputting one character at a time
    */
    while (*string != '\0')
    {
        data_ = *string;
        string ++;
    }
}
```

Note that the data members are tagged `volatile` to prevent compiler optimizations from skipping operations that you know you want to perform. The `volatile` keyword can be applied to declarations of objects, distinct from built-in data types, in C++, but I strongly recommend avoiding this. The member functions that are used on the `volatile` object have to be tagged `volatile`, and the only pointers allowed to access the object are pointers to the `volatile` type. These restrictions add to code complexity, and I have never seen a case in which declaring the object `volatile` was more advantageous than declaring the individual data members `volatile`.

The `TextDisplay` class restricts mixing ordinary data in with memory-mapped locations unless you know that the locations next to the memory-mapped device are mapped to RAM and are available. This is unlikely. If you must store such data, you should create a separate class that points at a `TextDisplay` class.

If you place classes in ROM using this mechanism, the compiler cannot guarantee that your code will never try to modify data members, but you can use the `const` keyword as described in the previous section to catch most cases. The initial assignment should then store the address in a pointer to `const`.

```
const Foo *fooPtr = new (location) Foo(a,b);
```

Objects created using placement `new` should never be deleted in the normal way. This is another instance in which it can be useful to disable the `delete` and `delete[]` operators. Often, such objects simply live until the program terminates. If you must run the destructor, it can be called explicitly

```
FooPtr->Foo::~Foo();
```

The placement `new` operator does not have to simply return the value passed in. The argument can define an area within memory in which the object should be located, or it can indicate which of several pools of memory should be used for the allocation (Saks, 1997).

7.4 Converted Examples

Most programming examples from previous chapters have been, to a greater or lesser extent, object-oriented. If you are C++-aware, you have probably noticed initialization functions that map to constructors and other mappings from C to C++. By comparing these programs with their C counterparts, I hope to establish the following points.

- Pointers to functions and casts can be eliminated in most cases.
- Using classes to protect data members leads to simpler expressions for accessing data members. Simpler expressions are more readable and less error-prone.

During the conversion from C to C++, I realized that many structure accesses crossed object boundaries. In some cases, I solved this by adding access functions, but in other cases, it was more appropriate to move the work from one class to another.

The benefit that C++ scales well as systems grow is not visible in these programs because they are too small as they stand now. You derive the real benefit of the TextDisplayModule class when you use several of them of different sizes, maybe within the same system and maybe on different projects.

7.4.1 Text Display Module

The C version of this program is presented in Chapter 2. In the original, I used an external declaration of the structure in the header file to name the TextDisplayManager, without giving away its internal description. In the C++ version, I put the class declaration in the header file. Although this means the class is visible in the other compilation units, only the public member functions can be called. Having the class visible in those compilation units is useful because the size of the objects is known. Thus, objects can be declared on the stack statically or declared on the heap with the new operator.

```
class TextDisplayManager
{
private:
    int lineLength_;
    int numberOfLines_;
    int numberOfLayers_;
    char ***layers_;
public:
    TextDisplayManager(int lineLength, int numberOfLines,
                       int numberOfLayers);
    ~TextDisplayManager();
    void refresh(void);
    void set(int lineNumber, int layer, char *string);
    void clear(int lineNumber, int layer);
    char * getSlot(int lineNumber, int layer);
    /*
      = 0 syntax means that this function is pure virtual.
      Any class deriving from this must define this function
      to allow instances to be created.
    */
    virtual void displayText(int lineNumber, const char *string)=0;
};
```

The displayText() function, which was a pointer to a function in the C example, is a pure virtual function here. When you use the class, you must first inherit from it in order to define this function. In the file tdmmain.cpp, I define such a class. The DosTDM is a TextDisplayManager that works with DOS screen. Obviously, for a real embedded system, you would define a class that works with an LCD or a dumb terminal on a serial port.

```
class DosTDM : public TextDisplayManager {
public:
    /* This defines a constructor that simply calls the parent
        constructor with the same arguments.
    */
    DosTDM(int lineLength, int numLines, int numLayers) :
            TextDisplayManager(lineLength, numLines, numLayers) {};
    void displayText (int lineNumber, const char *string);
};
void DosTDM::displayText(int lineNumber, const char *string)
{
    char buff[21];
    strcpy(buff, "                    ");
    strncpy(buff, string, min((int)strlen(string), 20));
    gotoxy(20, 22+lineNumber);
    puts(buff);
}
```

If you have only one text window, using inheritance is probably overkill because the `DosTDM` and `TextDisplayManager` can be merged into one class. However, this example demonstrates an important code reuse technique. By keeping the code that is common to all text displays in one class and the code specific to the hardware available in a particular case in a separate class, the common code can be moved to a different display with a minimum of changes, reducing the chances of introducing a bug.

For `TextDisplayManager`, the `TDMcreate()` function becomes the constructor. For this class, I also supplied a destructor so that if the class is declared on the stack, it will not result in a memory leak. The destructor frees up any resources allocated by the object.

```
TextDisplayManager::TextDisplayManager(int lineLength,
                    int numberOfLines, int numberOfLayers)
{
    int i; /* used to index the lines */
    int j; /* used to index the slots */
    /*
    These members could have been made constant, but we
    keep to more ordinary syntax for simplicity.
    */
    lineLength_ = lineLength;
    numberOfLines_ = numberOfLines;
    numberOfLayers_ = numberOfLayers;
```

```
/*
There is no need for assert here because C++ is defined
to raise an exception if new can not allocate the memory.
In this case it will cause the program to exit.
*/
layers_ = new char **[numberOfLines_];
for (i=0; i < numberOfLines_; i++)
{
    layers_[i] = new char *[numberOfLayers_];
    for (j=0; j < numberOfLayers_; j++)
    {
        layers_[i][j] = new char[lineLength_];
    }
}
/* Initialise each of the slots to an empty string */
for(i=0; i < numberOfLines_; i++)
{
    for(j=0; j < numberOfLayers_; j++)
    {
        layers_[i][j][0] = '\0';
    }
}
}
TextDisplayManager::~TextDisplayManager()
{
    for (int i=0; i < numberOfLines_; i++)
    {
        for (int j=0; j < numberOfLayers_; j++)
        {
            delete layers_[i][j];
        }
        delete layers_[i];
    }
    delete layers_;
}
```

For the `TDMset()` and `TDMclear()` functions (as with many others you see in this chapter), the TDM prefix has been dropped. Since the functions are now associated with a class, there is no risk of ambiguity or name clash with other functions using the names `set()` and `clear()`.

```
void TextDisplayManager::set(int lineNumber, int layer, char *string)
{
    char *startOfSlot;
    /*
    We could drop this assert if we were happy to lose characters
    off the right hand side of the display.
    */
    assert(strlen(string) <= lineLength_);
    startOfSlot = layers_[lineNumber][layer];
    strncpy(startOfSlot, string, lineLength_);
}
```

```
void TextDisplayManager::clear(int lineNumber, int layer)
{
    char *startOfSlot;
    startOfSlot = layers_[lineNumber][layer];
    *startOfSlot = '\0';
}
```

The last function, refresh(), provides no surprises. The interesting line is the call to displayText(), which is a virtual function call. Because no object is named with the call, it is assumed that the current object (the this object) is the one to use. In the example, the this object calls the displayText() function defined in the DosTDM class.

```
void TextDisplayManager::refresh(void)
{
    int i; /* used to index the lines */
    int j; /* used to index the layers */
    Boolean lineDone;
    char *startOfSlot;
    /* Loop through the lines */
    for(i=0; i < numberOfLines_; i++)
    {
        lineDone = FALSE;
        /* loop through the layers, searching for the first layer
           that has a nonempty string.
        */
        for(j=0; (j < numberOfLayers_) && !lineDone; j++)
        {
            startOfSlot = layers_[i][j];
            if (startOfSlot[0] != '\0')
            {
                lineDone = TRUE;
                displayText(i, startOfSlot);
            }
        }
        if (!lineDone)
        {
            /*
              There is nothing displayed on this line so the display
              blanks to cover anything that may have been on the
              line from a previous refresh. We know that startOfLayer
              is an empty string at the moment, and the display
              function will fill out the unused spaces with blanks.
            */
            displayText(i, startOfSlot);
        }
    }
}
```

So now this class can be used by creating an instance of the derived class, DosTDM, and calling the functions defined in the base class, TextDisplayManager.

```
TextDisplayManager * tdmPtr;
tdmPtr = new DosTDM(20, 2, 3);
tdmPtr->set(URGENT_LINE, URGENT_LAYER , "Urgent: OVERHEATING");
```

You can also use a DosTDM pointer to refer to the object. The following has exactly the same effect

```
DosTDM * tdmPtr;
tdmPtr = new DosTDM(20, 2, 3);
tdmPtr->set(URGENT_LINE, URGENT_LAYER , "Urgent: OVERHEATING");
```

7.4.2 multi *Graphics Demo*

The advantages of C++ are more apparent in the graphics world. There are many slight variations of graphical controls that can take advantage of inheritance. The amount of code that drives a graphics interface is, typically, an order of magnitude more than that required for a custom display. For this reason, the need for code reuse is greater. Graphics also suffer from more "creeping featurism" and late-change requests than other displays because there is so much flexibility in the functionality of a graphics display, long after the hardware design is set in stone.

The multi program used in Chapter 5 was rewritten in C++. I will not present the program in its entirety here, but I will extract the parts that illustrate differences between C and C++.

Shapes are a classic C++ example. I put the Button class in a separate file on the companion disk. This is not possible in the C version, in which all new shapes have to stay in the same file because it is the only place in the program that has access to the internals of the Drawable structure. Had the internals of the Drawable structure been visible in the header file, some encapsulation would have been lost, but the set of shapes might have been scattered throughout many files.

In the C version, the Drawable structure is contained within the various shapes. In C++, inheritance is more suitable. The base class is Drawable, which holds much the same data as the Drawable structure in the C example. The C++ class now holds many functions associated with that data. The Area structure is simple enough to leave as a structure and not convert to a class. If the program has to do a lot of manipulation of areas, such as merging areas or finding intersections, then forming an Area class is useful.

In Listing 7.2 the = 0 syntax indicates that the draw and erase functions are pure virtual. This means that you cannot create a `Drawable`, only a derived type that defines how to `draw()` and `erase()` itself.

Most of the functions are virtual, except for a few access functions. Sometimes, it is difficult to know which functions a derived type will need to overload. If there is any doubt, make it virtual. The overhead is slight, and it leaves the option open. If you will always have control over the source, it is not an issue. You can change the

Listing 7.2   An *Area* class can be a helpful addition to the *Area* structure for manipulation.

```
typedef struct
{
    int left;
    int top;
    int right;
    int bottom;
} Area;
/*
Forward declaration to allow the locateTarget function to be
defined in terms of Button.
*/
class Button;
/*
Forward declaration to allow pointers to container to be
returned by Drawable.
*/
class Container;
class Drawable {
private:
    Area area_;
    int color_;
    Container *parentPtr_;
    Drawable *nextContainedPtr_;
    Boolean dirty_;
    Drawable *nextDirtyDrawablePtr_;
    Container *oldParentPtr_;
    Area oldArea_;
    static unsigned char sallocBuffer[SALLOC_BUFFER_SIZE];
    static int sallocFree;
protected:
    void setArea(int left, int top, int right, int bottom);
    int getOldLeft(void);
    int getOldTop(void);
    int getOldRight(void);
    int getOldBottom(void);
```

prototype to virtual without making other changes to the function. It would have been reasonable for the language to make all functions virtual by default. But you would have had no way to avoid the small overhead of an extra level of indirection on each function call. This demonstrates that C++ is designed so that you do not pay for something unless you use it.

Listing 7.2 (continued)

```
public:
    Drawable(int left, int top, int right, int bottom);
    Drawable(void);
    virtual ~Drawable(void);
    void *operator new(size_t size);
    void operator delete(void *);
    Drawable *getNextContained(void);
    virtual void reparent(Container *newParent);
    Drawable *getNextDirtyDrawable(void);
    Container *getParent(void);
    Container *getOldParent(void);
    int getLeft(void);
    int getTop(void);
    int getRight(void);
    int getBottom(void);
    virtual void setPosition(int left, int top);
    virtual void move(int deltaX, int deltaY);
    /*
    Probably should be called setColor - but name might
    be confused with BGR's set color
    */
    virtual void setDrawableColor(int color);
    virtual int getDrawableColor(void);
    virtual void dirty(void);
    virtual void clearDirtyFlag(void);
    virtual Boolean isDirty(void);
    void addToContainedList(Drawable **containedListPtrPtr);
    void dropFromContainedList(Drawable *drawableToDrop);
    void dropFromDirtyList(Drawable *drawableToDrop);
    void updateOldData(void);
    /*
    Every descendent must define draw and erase, before an instance can
    be created.
    */
    virtual void draw(int offsetX, int offsetY) = 0;
    virtual void erase(int offsetX, int offsetY) = 0;
    virtual Button *locateTarget(int x, int y);
};
```

Several access functions that allow other classes access to the stored `area_` and `oldArea_` are supplied. These are good candidates for in-lining, but for readability I left them out-of-line.

The `oldArea_` is of concern only to the derived types because it is used by the `erase()` function to calculate the position of the old image. The access functions for that data member are `protected`. The `protected` status means that derived types can call it, but others cannot.

The constructor initializes all the data members, some of them according to the arguments passed in. There is also a default constructor that takes no arguments. This is useful if some derived types do not have the arguments available at the start of their constructor. In those cases, the default constructor is implicitly called. The `Area` can be set later in the derived types constructor using the `setArea()` function.

There is no function equivalent to the destructor in the C code, and I never destroy any of the shapes in the sample code on the companion disk. However, I added a destructor here to illustrate some typical tidying up. There are no resources allocated in the constructor, but by the time the object is destroyed, it may be on several linked lists. You should remove the object from those lists to prevent an illegal access the next time the list is traversed. Removing the object from its current `Container` removes it from the contained list. However, the act of removing it from the `Container` puts the object on the dirty list. So it must be removed from there also. Similarly, the `Container` destructor must ensure that its children do not remain pointing at the deleted `Container` as a parent.

```
Drawable::Drawable(int left, int top, int right, int bottom)
{
    color_ = BLACK;
    area_.left = left;
    area_.top = top;
    area_.right = right;
    area_.bottom = bottom;
    parentPtr_ = NULL;
    nextContainedPtr_ = NULL;


/*
The object is not initially dirty. It will be marked dirty when
it is added to a container.
*/
    nextDirtyDrawablePtr_ = NULL;
    dirty_ = FALSE;
    oldParentPtr_ = NULL;
}
```

```
/*
This is the default constructor. It is called if no arguments are
passed to the Drawable class when the derived class is constructed.
*/
Drawable::Drawable(void)
{
    color_ = BLACK;
    area_.left = 0;
    area_.top = 0;
    area_.right = 0;
    area_.bottom = 0;
    parentPtr_ = NULL;
    nextContainedPtr_ = NULL;
/*
The object is not initially dirty. It will be marked dirty when
it is added to a container.
*/
    nextDirtyDrawablePtr_ = NULL;
    dirty_ = FALSE;
    oldParentPtr_ = NULL;
}
/*
This is the destructor for Drawable. It has no resources to free,
but it must remove the Drawable from any lists it may be on
that may be traversed later.
*/
Drawable::~Drawable(void)
{
    if (parentPtr_ != NULL)
    {
        parentPtr_->remove(this);
    }
    G_display.dropFromDirtyList(this);
}
```

The other classes inherit from `Drawable`, each adding whatever data they require and overloading the `draw()` and `erase()` functions. `Box`, `Circle`, `Line`, and `Text` are fairly straightforward.

```
class Box : public Drawable
{
private:
    Boolean filled_;
    int fillColor_;
    Boolean oldFilled_;
```

```cpp
public:
    Box(int left, int top, int right, int bottom);
    virtual void fill(int color);
    virtual void draw(int offsetX, int offsetY);
    virtual void erase(int offsetX, int offsetY);
};
class Circle : public Drawable
{
public:
    Circle(int centerX, int centerY, int radius);
    virtual void draw(int offsetX, int offsetY);
    virtual void erase(int offsetX, int offsetY);
};
class Line : public Drawable
{
private:
    int x1_;
    int y1_;
    int x2_;
    int y2_;
    int oldX1_;
    int oldY1_;
    int oldX2_;
    int oldY2_;
public:
    Line(int left, int top, int right, int bottom);
    virtual void draw(int offsetX, int offsetY);
    virtual void erase(int offsetX, int offsetY);
};
class Text : public Drawable
{
private:
    char *string_;
public:
    Text(int left, int top, char *string);
    virtual void draw(int offsetX, int offsetY);
    virtual void erase(int offsetX, int offsetY);
};
```

The constructor of each one passes its arguments directly up to the Drawable constructor, with the exception of Text, which is different because the full area can be established only by checking the width and height of the text in the string supplied.

```
Box::Box(int left, int top, int right, int bottom)
    : Drawable(left, top, right, bottom)
{
    filled_ = FALSE;
    oldFilled_ = FALSE;
}
Circle::Circle(int centerX, int centerY, int radius)
      : Drawable (centerX - radius, centerY - radius,
                  centerX + radius, centerY + radius)
{
    /* Nothing to do */
}
Line::Line(int x1, int y1, int x2, int y2)
    : Drawable(min(x1, x2), min(y1, y2), max(x1, x2), max(y1, y2))
{
    x1_ = x1;
    y1_ = y1;
    x2_ = x2;
    y2_ = y2;
}
Text::Text(int left, int top, char *string)
{
    int right;
    int bottom;
#ifdef USING_BGI
    right = left + textwidth(string);
    bottom = top + textheight(string);
#endif
    setArea(left, top, right, bottom);
    string_ = string;
}
```

The new operator is overloaded to allocate space for each object. It is only overloaded in the Drawable class, and it is then inherited in all the derived classes. The algorithm used is exactly the same as in the C version. The static array used as the pool of memory is declared within the Drawable class to prevent external access. As a static member of the class, there is only one instance of this array shared among all instances of Drawable. The delete is overloaded also to prevent memory allocated by the overloaded new operator from being returned to the global allocation scheme.

```
void *Drawable::operator new(size_t size)
{
    void *nextBlock;
    assert(sallocFree + size <= SALLOC_BUFFER_SIZE);
    nextBlock = &sallocBuffer[sallocFree];
    sallocFree += size;
    return nextBlock;
}
```

7.4.2.1 The Display Class

Most functions make simple accesses to the member data, calling the dirty() function if any changes are made that require a refresh of the object. The mechanism for redrawing based on a list of dirty objects is exactly the same as in the C version. Virtual functions have eliminated the need to store the type data member in Drawable. The switch statement in the drawableDraw() function in the C version disappears. Instead, the pure virtual Drawable::draw() function is called, and this calls the draw function of the appropriate derived class.

The refresh() function lives inside the Display class. This class also holds the list of dirty objects. You declare the one required instance of this class in shapes.hh so that the event loop can access it when refreshing is required. Note that this class is not in the hierarchy with Drawable as the base class, because Display itself does not have an image or position on the screen. The class is defined as the following.

```
class Display
{
private:
    Drawable *dirtyListPtr_;
public:
    Display(void);
    Drawable *addToDirtyList(Drawable *drawablePtr);
    void dropFromDirtyList(Drawable *drawablePtr);
    void refresh(void);
};
extern Display G_display;
```

The refresh function has much the same form as before: looping through the dirty list, erasing the old image of each object, and drawing the new one.

7.4.2.2 The Container Class

Now that you have done the transition to C++, the Container class becomes much more interesting. The Container itself holds a chain of contained Drawables and inherits from Drawable to manage its own Area. However, you can now inherit from Container to form the three main dialogs defined in the multi.cpp file. First, look at the Container class and then at its derived classes.

```
class Container: public Drawable
{
    Drawable *containedListPtr_;
/*
The absolute location is maintained to optimize drawing the
contained objects.
*/
    int absoluteLeft_;
    int absoluteTop_;
    Boolean visible_;
```

```cpp
public:
    Container(int left, int top, int right, int bottom);
    ~Container(void);
    virtual int getAbsoluteLeft(void);
    virtual int getAbsoluteTop(void);
    virtual void add(Drawable *newDrawablePtr);
    virtual void remove(Drawable *drawablePtr);
    virtual Button *locateTarget(int x, int y);
    virtual Boolean isVisible(void);
    virtual void reparent(Container *newParent);
    virtual void draw(int offsetX, int offsetY);
    virtual void erase(int offsetX, int offsetY);
    static Container *createRoot(int width, int height);
};
```

The add() and remove() functions simply insert and delete items from the containedListPtr_ linked list. However, these functions can no longer directly manipulate the parent_ data member of Drawable, so the reparent() function is added to Drawable to allow changing of the parent_ field. The reparent() function is overloaded in Container because it is necessary to update the visible_ flag for the Container and for all sub-Containers, so although reparent() in Drawable is simple, the overloaded version in Container must loop through all its children. Just like the C version, all Containers must maintain a visible_ flag so that they know if they are connected to the root Container.

```cpp
void Drawable::reparent(Container *newParentPtr)
{
    parentPtr_ = newParentPtr;
}


void Container::reparent(Container *newParentPtr)
{
    Drawable *iterator;
    if ((newParentPtr == NULL) || !(newParentPtr->isVisible()))
    {
        visible_ = FALSE;
    }
    else
    {
        /*
        The visible flag must be set to TRUE because it is now
        connected to the root
        */
        visible_ = TRUE;
    }
```

```
    /* Now call the Drawable to update the private parent data. */
    Drawable::reparent(newParentPtr);
    /*
    Call reparent on any children to update their visible flags.
    */
    for(iterator = containedListPtr_; iterator != NULL;
                iterator = iterator->getNextContained())
    {
        /*
        This call to reparent does not change the parent,
        but it updates the visible flag appropriately if the
        child is another container.
        */
        iterator->reparent(this);
    }
}
```

Making a function recursively descend though a containment hierarchy and making the recursion enter different functions depending on the type of the object is a common technique. The `locateTarget()` function uses a similar technique to identify the Button currently under the pointer. If the nested containers are very deep, you may want to convert this recursion to a loop. This improves performance, but more importantly, it limits the stack space required.

7.4.2.3 Inheriting from Container

Now that Container is defined, the dialogs defined at the application level can inherit from Container. For example, the graphs dialog can be as follows.

```
class GraphsDialog : public Container
{
private:
    Trace *leftTracePtr_;
    Trace *rightTracePtr_;
    Boolean leftTraceLive_;
    Boolean rightTraceLive_;
    Button *leftButtonPtr_;
    Button *rightButtonPtr_;
public:
    GraphsDialog(void);
    Boolean handleKey(char key);
    void handleSampleEvent(SampleEvent *samplePtr);
};
```

Instead of adding the buttons and traces to a separate `Container`, the `GraphsDialog` simply adds the contents to itself. Real gains in the simplicity of the C++ syntax are now evident. Consider the following few lines from the C version.

```
axisPtr = lineCreate(29, 100, 29, 300);
containerAddTo(GS_graphs.containerPtr, GET_DRAWABLE(axisPtr));
```

This code created a line that is one axis for a graph, and then it added it to the `Container`. In C++ it becomes the following.

```
axisPtr = new Line(25, 100, 29, 100);
add(axisPtr);
```

Creating the line is just as simple, although now you use the new operator. Adding the line to the `Container` is reduced to a simple function call. You do not need to specify which `Container` is the object of the `add()` function, because this code is in the constructor of the `GraphsDialog`. And `GraphsDialog` is a `Container`, so the line is added to the `GraphsDialog` itself. The `GET_DRAWABLE` macro is no longer necessary, because any `Drawable` can be passed to the `add()` function and `Line` is a `Drawable`.

Simplifying the code in this way makes it more readable and, therefore, less error-prone. However, be warned that actions that are no longer explicit can catch you if you make a wrong assumption.

Now that I have discussed the different types of `Container`s, I will examine the inheritance hierarchy, shown in Figure 7.4. Compare this with the tree in Figure 5.4, which shows the containment hierarchy at a particular stage in the program. Figure 5.4 is just as applicable to the C++ version as it is to the C version, although the nodes

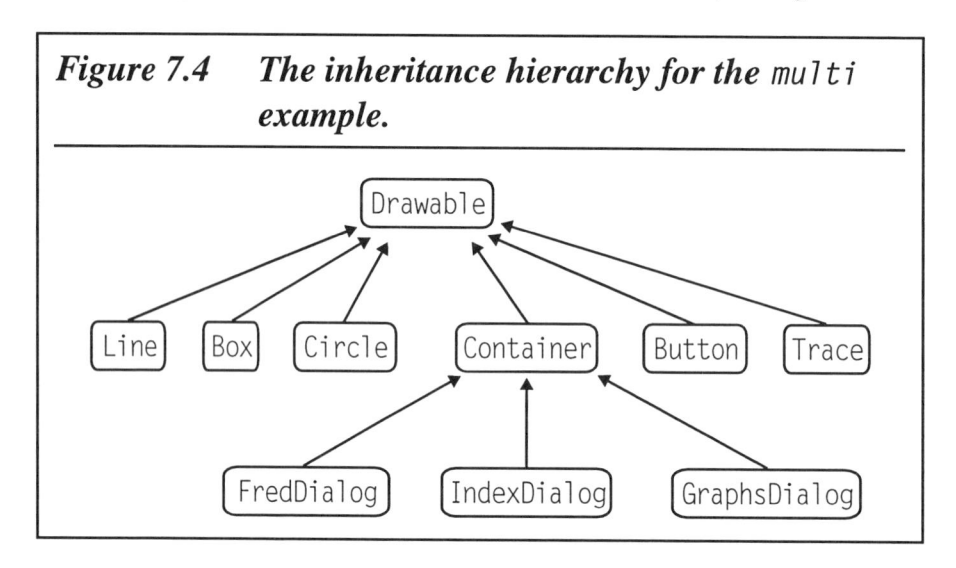

Figure 7.4 The inheritance hierarchy for the `multi` *example.*

of the tree are now C++ objects rather than structures. The containment hierarchy is dynamic, and it changes at run time as objects are added and removed. The inheritance hierarchy, on the other hand, is static; it freezes at compile time.

7.4.2.4 Lists of Objects

The lists in this example are threaded through the objects by having a pointer contained within the object for each list that may include it. The contained list for managing Containers and the dirty list for refreshing are the two lists in this example, and the links are maintained in the nextContainedPtr_ and nextDirtyDrawablePtr_ data members of Drawable. This is not a clean or maintainable way of managing lists. In mainstream C++, a template (possibly from the standard template library) can be used to generate a list that handles objects of type Drawable. A separate list object would contain pointers to all objects on the list. In this example, I avoided this approach for several reasons. A separate list would have to allocate space for its pointers dynamically, which is not desirable in embedded systems. Alternatively, the list could have a fixed maximum size. If each Container contained a list large enough to handle the most Drawables it could ever contain, much space would be wasted. Few Containers would use up more than a fraction of the space set aside for those pointers. You would also face the risk that some screen configuration might exceed that maximum when you least expect it.

For these reasons, I use lists that thread themselves through the existing classes. The size of each Drawable is a little bigger, but the size increase is fixed and you know there will always be a link when you require it. A pleasant side effect is that one Drawable can never appear on more than one contained list and one dirty list. This is a case in which the design of the data is simpler. It does mean that the code is not as elegant as it might be, because much of the iterating through the linked lists is in the Drawable class, which should really be concerned only with how to draw itself and manage its screen location. If your project requires many lists for other purposes, perhaps you should build a generic list. With only two lists, though, this example program suffers some code repetition in the handling of the lists.

I will now discuss how you remove a contained Drawable from a Container. The remove() member function of Container is called to handle the special case of the child that is to be removed, being the first one on the list. If it is not the special case, control is passed to the Drawable::dropFromContainedList() function, because the containedListPtr_ data member that is owned by Drawable must be manipulated. Within that function the list is iterated, searching for the item to be removed.

```
void Container::remove(Drawable *childPtr)
{
    Drawable *iterator;
    assert(childPtr != NULL);
/*
If the child to be dropped is the first on the list, it is
handled slightly differently because the value of
containedListPtr_ has to change.
*/
    if (containedListPtr_ == childPtr)
    {
        containedListPtr_ = childPtr->getNextContained();
    }
    else
    {
        containedListPtr_->dropFromContainedList(childPtr);
    }
    childPtr->reparent(NULL);
    childPtr->dirty();
}
void Drawable::dropFromContainedList(Drawable *drawableToDropPtr)
{
    Drawable *drawablePtr;
    /*
    If the next one on the list is the one to drop, change the
    nextDirtyDrawablePtr_ to skip over it. Otherwise, ask the next
    guy on the chain to do the same check.
    */
    for(drawablePtr = this; drawablePtr != NULL;
                drawablePtr = drawablePtr->nextContainedPtr_)
    {
        if (drawablePtr->nextContainedPtr_ == drawableToDropPtr)
        {
            drawablePtr->nextContainedPtr_ =
                drawablePtr->nextContainedPtr_->nextContainedPtr_;
            return;
        }
    }
}
```

You should note that the Drawable::dropFromContainedList() function accesses the private data of its own object and all of the other objects on the list. Although C++ allows access to the private data of another object of the same class, this technique is not particularly object oriented. However, the alternative is to call a function on each object in the list until you find the object to delete. Because you do not know how long your lists will be, this could be time consuming. This is another case in which I neglect object-oriented practices in order to preserve performance. The sacrifice in encapsulation is mitigated by the fact that the code for manipulating all the `Drawables` on the list is still within the `Drawable` class, so changes to this mechanism are still local to the class.

7.5 Conclusion

So what have you learned about C++? Many features gain real value only as a program gets bigger. Many features are oriented toward putting a little extra effort into the class design so that programming is easier for class users. For example, if you have many constructors or many ways of accessing the data within a class, the user of the class will have more options. However, it requires extra work by the creator of the class. This makes sense if you use the class in many places, but the return on investment may not be as great if you use the class in one small program.

Learning C++ is not trivial. Many programmers go through a full project of many months before they feel comfortable with it. So it is difficult to get immediate return on investment. Later projects, especially if they are in a similar application domain, provide greater return as the knowledge and some of the previous project's classes can be reused.

Restricting the C++ features you use significantly reduces both the learning curve and the difficulty of bringing new team members up to speed. I will not say "never use such-and-such a feature"; different designs have different demands. However, you should avoid making heavy use of every feature.

C++ is far from a perfect language, but it will be with us for a long time, and it has much to offer. If you decide that C++ is not appropriate for a project, then I hope this chapter has given you enough information to make your decision an informed one, not a guess based on C++ myths.

Chapter 8

The Design Process

So how does the user interface fit in with the rest of the design process? The overall design process may not influence the color of your buttons or the key-debouncing algorithm. However, it does influence your choice of prototyping tools and the way you manage user interface requirements. This chapter explores some of the issues raised in making the user interface work with the rest of the design process, distinct from the rest of the product.

I avoid discussing the many structured and formal methods of analysis and design of requirements. Why? First, in my experience, very strict methods for analysis cause more problems than they solve. This may not be true in general, but it is most certainly still open to debate. Second, my goal is to explain what you need to develop a user interface. You can decide after reading this chapter if your needs in the user interface area can be met by the requirements analysis method that you are using. You should note that in some closely monitored cleanroom projects (Selby et al., 1987), developers felt that the user interface suffered because they were isolated from the final system by the strictness of the programming methodology. The cleanroom methodology does not allow the developer to compile or run the code himself. Because he cannot run the code, he gains little experience of the user interface as the product advances. Other strict methodologies may suffer from a similar effect.

You must decide what the defining document of the user interface is to be. There are three main approaches. The first approach is a design document that defines (through English or pictures) how the user interface should perform. This becomes the source document for the software, the tests, and the user manual. This approach allows work on the user manual to proceed without having a complete version of the software.

You can also make the user manual the source document. This works if the project is small and the user manual contains enough detail. There will be conflicts, however, between the level of detail required for the user and what must be recorded to completely specify the user interface.

A third approach is to build a prototype and declare that the prototype defines how the final product should work. On a small project, you may demonstrate the finished product or a convincing prototype to a curious nonprogrammer and tell him, "Play with it and see what happens." Many programmers would love to follow this route. Allow the code to define the interface, and you can always run the system to find out what it is defined to do. This cowboy approach simply does not work on a large project. Building the prototype is not a bad idea, but allowing it to be the ultimate definition of user interface requirements is dangerous. The prototyping environment may not be able to represent some of the properties necessary in the final product. Mistakes in the prototype may be defined as features and copied in the final implementation.

For the remainder of this section, I will consider the first option — a design document that defines how the user interface should perform. This option is relevant for almost all nontrivial projects.

Many projects maintain a document tree. At a high level, the user interface requirements may dictate only that the user can control the device in certain ways. The details of the interaction should not be shown. If you work in a disciplined environment, the lowest level documents describe the software design. The description includes data structures, function names, and algorithms. In between, it is important that the user interface requirements document is written at a low enough level to define the interactions that take place without assuming the reader has programming knowledge. Testers, technical writers, and marketing people may not understand programming, but they all want a good definition of what happens if you press the big red button. Try to force all information at this level into a single document. Do not allow it to spread out across several documents at different levels. Also, avoid non-user-related issues in the same document. Many people may be involved in the user interface who are not involved in the rest of the project, and a single document used as a common resource for those people is valuable.

Figure 8.1 shows how the user interface requirements document relates to the rest of the process. Obviously, this diagram varies depending on the types of documents dictated by your process. The principle, however, remains the same. You should derive the tests from requirements and not from prototypes or the real system. If the tests are based on the implementation or the prototype, the tests will guarantee only that the implementation is internally consistent or that the implementation is consistent with the prototype. If tests are based on the prototype, you may miss some requirements completely because they were too difficult to implement in the prototype.

Although most usability feedback comes from the prototype, some inevitably comes from the real system, because there will be features that the prototype does not fully implement. Also, the prototype cannot always be deployed as effectively as the real system. You can prototype a cash register on a computer screen, but you cannot put that system in place in the canteen at your workplace. You may, however, be able to do so with a fully working prototype.

One of the user interface requirements document's most important roles is as a repository of decisions that are made during meetings and brain-storming sessions. A document can be updated quickly, although there might be a few weeks' lag before the same feature is implemented in the code. In the intervening days, other work can be done — documenting, testing, or programming. The person doing that work has no reference to the latest plan unless he is involved in all discussions of new features, which may not be feasible or desirable on a large project. A defining document gives everyone involved a single point of reference. If the document cannot be revised after every decision, you should keep one marked-up master copy that contains notes, possibly hand-written, describing the changes that will be included in the next revision. This copy might exist on paper or as an on-line document. Dedicated requirements tracking tools are also available.

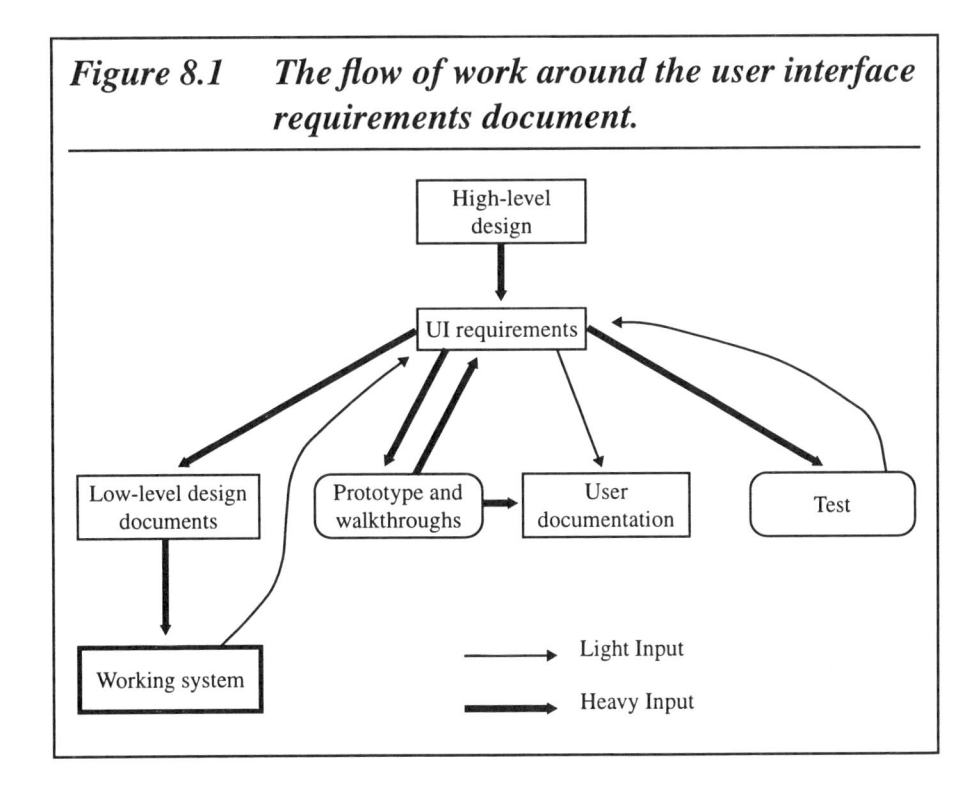

Figure 8.1   ***The flow of work around the user interface requirements document.***

8.1 Writing Requirements

A formal syntax to describe how the interactions work seems tempting from the engineer's point of view, but it can alienate the less technical people involved in requirements definition. Thimbleby (1990) explores this issue in *User Interface Design*. The application domain offers alternative formal methods of describing changes caused by user actions. For example, chemical reactions can be described using the chemical reaction equations common to that industry. A chess game can use the standard algebraic notation that allows moves to be described in shorthand. In more general cases, pseudocode can be useful.

For something as simple as using arrow keys you can use:

```
The UP key shall have the following effect on the position:
if (x < x_max)
position_x = position_x + 1;
else
position_x = x_max;
```

This pseudocode describes the behavior of the Up key more precisely than an English sentence could. However, the pseudocode must remain at a level at which the nonengineering members of the team can interpret it. They need to review the requirement for appropriateness or convert it into user documentation. Avoid C-specific idioms, such as `position_x ++;`, which are obvious to the programmer, but not so obvious to other readers.

Make requirements readable by the target customers.

The level of detail needed for a requirements document varies widely. On a graphics project, stating that the program is based on a specific third-party library implies that the low-level details of the interactions must conform to that library. No further explanation should be necessary for features described in the library's documentation.

Structure the requirements to avoid repetition. Consider the requirement that button A will cause a key click sound followed by a certain sequence of actions. Another requirement is that button B will cause a key click sound followed by a different series of actions. It is better to have a single requirement stating that buttons cause a key click sound. This requirement can list the exceptions if there are any. By making more general requirements, the document maps more closely to the code that implements it. Changing requirements is easier and safer because you do not visit as many places in the document. There are fewer chances for requirements that are far apart in the document to describe actions that are inconsistent with one another.

8.1.1 Walkthroughs

A particular interaction or series of interactions is difficult to visualize with a user interface requirements document. The reader needs to know which general rules may be applied at any time, and these rules may be scattered throughout the document.

In some cases, you may need to examine the system response to a sequence of events in a walkthrough of the interactions. You can use this to evaluate the system from a usability point of view and to give the team a better feel for how the final system will work. You can take a group through a walkthrough in several ways: a simple verbal description of the sequence; a series of pictures showing what the user interface looks like after each action; or best of all, a demonstration of a prototype. These exercises help define the requirements, but be wary of using the same format to document requirements. The number of walkthroughs required to completely define a user interface grows exponentially as the interface grows. This leaves you with a huge maintenance problem if you need to rework walkthroughs. By their nature, each step of the walkthroughs contain many details. These details may contradict each other: details that are not significant in a walkthrough may be inserted with less care than the details the walkthrough is trying to demonstrate. A walkthrough that demonstrates a mode change may have buttons of a particular size. If you decide to change the button size everywhere, does that mean that all walkthroughs should be updated? Do you just update the walkthrough that defines button appearance? You can outline the aspects a walkthrough aims to define, but it is not a clean approach. Walkthroughs should test and derive requirements. They are not a good method of documenting the system requirements.

When you use walkthroughs to explore new ideas, pen and paper can be more flexible than computer graphics. When you are choosing from a set of well-defined components, cut out colored pieces of paper that represent components, and draw the button, numeric display, or dial. Rearrange those pieces of paper until you find a configuration you like. Cutouts of text messages are easy to add and remove; you may not be able to erase easily on paper. If this sounds like reducing design to a children's playground, all the better — imagining new ways of doing things should be fun. If the medium is transient, you can use an instant camera to freeze some promising ideas for later review.

8.1.2 Testability

When you define requirements, ask yourself, "Is this testable?" The requirement that all user requests should be handled in a timely manner is not testable because "timely" is a subjective measurement. You can decide that all keys must be processed within some time period. Or, if you know that certain keys always take a long time to process, define an average, or state the exceptions.

Ease of use is difficult to test. Although it is a valid aspiration, you should not state it as a requirement unless you are prepared to define a test of the property based on observations of sample users and a measurement of their performance.

8.1.3 Assumptions About the User

Avoid making assumptions about the user in the requirement definition. Consider the following requirement.

"When there is no CD present and the tray is closed, pressing the open/close button on the CD player will open the CD tray. After the user has placed a CD on the tray, pressing the open/close button again will close the tray. The CD is now available to be played."

You can assume that there is a separate requirement to cover the case in which the open/close button is pressed while there is a CD present. That requirement addresses issues such as stopping play and other details not addressed here.

This requirement assumes that the user will place a CD on the tray. The user may not do this. He may close the tray again. He may walk away and leave the tray open indefinitely. It is legitimate to make assumptions about what the device will do, because another requirement can define the way the device must behave. Device requirement failures imply that the device has failed and requires special attention. The same is not true of the user, although there are programmers who might try to convince you that users who do not conform to their program expectations have, in fact, malfunctioned and should be fixed.

Assumptions about the user are dangerous because they may lead you away from considering alternative actions. The programmer needs to give equal attention to the case in which the tray is closed without a new disk, as well as to the case in which there is a new disk present. The earlier requirement would be better reworded as:

"When there is no CD present, pressing the open/close button on the CD player will open the CD tray. Pressing the open/close button again will close the tray. If a CD is detected on the tray, it is now available to be played."

In other cases, the system cannot directly detect the user's action, so it is even more important to allow for alternatives. For a safety-critical system, any user actions that might jeopardize the performance of the device should be contained in the hazards analysis. See Chapter 6 for a discussion of hazards analysis.

Make no assumptions about the user. You are defining the device, not the user.

Walkthroughs and assumptions about the user are more appropriate in the user manual, where the objective is to teach rather than to define.

8.1.4 Learnability

A user interface requirement should always be *learnable*. It is not only engineers and testers who must learn each requirement. If a user is to become fully competent with the device, he must learn each user interface requirement. He may learn them from use and experience rather than from a requirements document. If a requirement is so hidden that the user is never aware of it, that requirement is, by definition, not a user interface requirement. If a rule is so complex that no user can understand it, then sometimes the user will not be able to predict the system behavior.

Take the example of a set of traffic lights. The only requirements the designers consider external are that green means go, red means stop, and yellow means prepare to stop. The designer may assume the driver will not want to learn anything beyond these simple rules. However, regular drivers may notice the synchronization of sets of lights at successive junctions. Assume the following rule for synchronization: A car leaving one junction when the light turns green and traveling at 30mph will arrive at the second junction as the light turns green. The regular driver — the expert user — learns this rule after a little trial and error. He can then use the system to his best advantage. You must also be careful that the driver does not learn something he is not supposed to know. Some drivers may discover that they can synchronize with the lights by traveling at twice the design speed! Designers have to introduce offsets to prevent this.

What if the designer decides that drivers travel slower on narrow roads and faster at certain times of the day? By programming in these attributes, the lights may synchronize better with the average driver. However, the regular driver, who tries to learn the system is now caught out. The new system's rules are too complex to work out. Even if they are given to him, there are too many combinations of roads and times of day for the driver to pick his speed for different parts of the journey. The more sophisticated analysis of the user's behavior yields a system that is less learnable and that can frustrate the expert user. In this case, learnability is traded off for better throughput from the system as a whole. This may be the correct design choice in this case, but you should realize where the sacrifices are made to achieve this overall improvement.

Strive to make requirements learnable.

You may not be able to construct requirements that are learnable, testable, and that make no assumptions about the user; however, these are desirable properties. If you create a requirement that lacks them, be aware of the drawbacks.

8.1.5 Maintaining a Glossary

A glossary, if enforced from the beginning of a project, is an ideal way to protect the namespace of the English language used by the team members. Each project has a set of words that are over-used. Is the *screen* the physical display containing a cathode ray tube, or is it one of many display configurations that appear on the cathode ray tube? Is the *indicator* light the one that tells me that the car will turn left, or does it include the small lights on the dashboard that indicate low oil or high engine temperature? Does *IPC* mean interprocessor communication (between CPUs) or interprocess communication (with one CPU running many processes)? If the system is *restarted*, was the power cycled by the user, or was the CPU reset without losing power?

Sometimes, if more than one interested party wants to use the same word or phrase, they can all be convinced to find new ones. This may sound like manipulation of the English language to control the way people think, as Orwell describes in the novel *1984*. I am not suggesting anything so draconian. It does not matter if the glossary follows the way people talk, as long as there is an agreed interpretation of meanings in documents that describe the product. If you must have per-document glossaries, use a subset of the project-wide glossary.

Most glossaries cover the meaning of acronyms and words that are specific to the product. Usually, the most troublesome words are the normal English words that are omitted from the glossary because they are considered obvious. Everyone knows what a button is — until you have buttons drawn on a graphics screen and mechanical buttons both.

You may wish to distinguish the words you add to the user's vocabulary. You can use the words in user documentation and present them to the user on a display or on a label on the user interface. The terms not specified as part of the user's view of the product are for internal use only.

8.2 Iterations Through the Interface

Development of a user interface is not amenable to a single pass. All steps will be iterated until either the interface is stable or management cries, "Ship it!" During development, a feedback loop exists between the developers and the users or the marketing function, which may represent the users. The developers design an interface or part of an interface, which is then presented to the users in paper form, as a partially working prototype or as a fully functioning system. The user's reaction to this interface is evaluated, which leads to changed requirements and further trials. This feedback loop generally decays over time as changes become minor and do not lead to further changes.

Unfortunately, sometimes the changes that occur with each iteration are so large that they always lead to further changes. Often, a change intended to have only a minor impact can upset the overall consistency of the interface, leading to compromises in unrelated areas. When this happens, you should reevaluate the initial change.

In the early stages, the rate of change of requirements can be high. One reason is that the interactions may be so complex that many flaws and inconsistencies in a specification are not discovered until it is implemented. One part of the specification may say, "Under condition A, turn on light X." Another part it may say, "Under condition B, turn off light X." During implementation, it may transpire that A and B can simultaneously be true. So now what do you do with light X? Also, when interactions are specified, the designer may rely on his imagination or on prototypes that do not accurately reflect the real system. The real thing might not live up to expectations.

These changes contribute to the iterations of the feedback loop. Damping this feedback loop to decay rapidly is the key to meeting deadlines. Usually the engineers are under deadline pressure, and they must make sure all changes are justified or suffer the consequences. Refusing new changes after a certain date sounds simple, but user interfaces suffer from a political problem not suffered by other parts of the system. For many devices, anyone who looks at the device understands the user interface. Later in the project, more and more people get to see the product. As the project nears launch, the company's sales force and service organization get involved. Potential first customers are shown a product that was kept a closely guarded trade secret up to this point. Managers higher up the food chain show an interest, hoping to bask in some of the reflected glory should the product prove successful. As these parties get to look at the new toy, many want to make suggestions. The mechanical and electronic design will not change, because the associated tooling costs are so high. In any case, the design issues will not be accessible to many people in the new-found audience. However, the user interface is accessible to all. Everyone feels he can make an input on the workings of the interface. There will be requests for an extra message, to flash a LED at a different rate, to allow the user to select function A even when they are in mode X. Resisting the late flood of suggestions and requests is difficult when some are genuinely good and some come from influential customers or managers. There will be hard decisions to be made about what is needed to ship the product and what can be held over to a later release. Maintaining a formal wish list for future releases is a good way to manage late suggestions without losing time on the schedule or valuable input.

A warning about late-breaking changes in the project's requirements: change can be bad for the structure of the software. Although the interface may be incrementally improved, each round of change can compromise the internal design as the software is coerced into doing something not originally envisaged. All too often, changes are implemented with a hack, only to come back to haunt you. The coder rationalizes these hacks by planning to code it properly after the user interface stabilizes. I will leave it as an exercise to figure out how often that happens.

You can try out new versions of the user interface in a special build of the software or in a prototype, with no changes committed to the master sources unless you decide to accept these changes. This process does not always work, because all the new features must be seen together; some survive, others do not. Some features survive the first round, only to be shot down during a future round. Some iterations can be avoided by making good use of prototypes, but you can rarely freeze the user interface requirements the day the coding starts.

8.2.1 Test Team's Role

The test team are probably the first to see a fully built system and to use it in realistic circumstances. The previous prototypes and mock-ups cover many possible interactions, but they tend to happen in a controlled environment and with a limited set of tasks. If the interactions are with a mock-up, they may miss out on some interactions that can happen with a real system. Consider a microwave front panel. You can perform usability tests with a prototype displayed on a desktop computer or with a front panel and a nonfunctioning oven. The usability tests may establish a good dialog for changing time and power settings. When the test team uses real ovens, they may find that for some settings the food or containers burn the user's hands when he tries to remove food from the oven. If the door latch is controlled from software, the test team may recommend adding a short time delay to the door-opening mechanism after the food is cooked, to reduce the chance of a serious hand burn. This input can not emerge from the usability trials, because the trials do not cover all aspects of use. This is partly the trial's fault, but you can never make the trials completely realistic. Even after the most exhaustive usability trials, the test team will still spend more hours on the machine than will any of the sample users.

The test team also spends more time using the complete system than do the developers, who tend to focus on their own areas. Because the testers did not create the system, they have a more objective view of the product than the developers. The test team is technically proficient and familiar with the product domain. They recognize what can and cannot be changed, and they can describe the problems concisely. This greatly improves the signal-to-noise ratio when the feedback is filtered by the developers. Feedback from sample users, who are, by definition, less familiar with the product, includes many suggestions that cannot be used.

For these reasons, the test team should be given the mandate to discover usability problems. Instructing the test team to report only nonconformance to requirements may get the product out the door faster, but chances of it being the right product are greatly diminished. You must accept that some usability issues cannot be foreseen at the requirements stage. It is better to allow the test team to discover these problems early than for the customer to discover them later.

8.3 Simulations and Prototypes

For purposes of this section, I use the word prototype to describe something that attempts to look and act like the real user interface. Manipulating the controls may not perform the expected actions. You might prototype a car's dashboard on the screen of a computer but operating the accelerator with the mouse will not cause the workstation to leave the room at 60mph. On the other hand, a simulation of the user interface acts like the user interface as far as the rest of the application is concerned. The simulation may not bear any visual resemblance to the actual user interface. The simulation can be a set of scripts and input files or a C program that puts the application through the same steps that occur if the user provides certain input. The simulation also allows output functions to be called, although there is no output hardware available. These outputs may be recorded for debugging purposes. This definition of simulations excludes flight simulators, which simulate flying from the pilot's point of view and which, by my definitions, are prototypes. The simulators discussed here simulate the user, or the user interface from the application's point of view.

Prototype and simulation tools are useful at different phases of the design cycle. The prototype can define and refine the user interface requirements, the simulation can test and debug the software, possibly before user interface hardware is available.

Later in the design cycle when the actual system is available, both tools become redundant. Occasionally, the simulator still provides a better debugging tool if it allows the tester to generate user input that is too time-consuming for a person to input, or if it is necessary to reproduce an exact sequence of events several times. A human operator may not be able to repeat a sequence with the exact same time intervals between actions. The prototype will not be used to demonstrate the latest version of the system once the real system is available, but it may still be useful for quickly trying a refinement that would be too time-consuming to put into the real system.

8.3.1 Simulations

Simulations have a checkered history in the embedded world. In most cases, the timing constraints or the hardware cannot be simulated for a lower cost than building the system itself. Real hardware, even if not fully debugged, is preferable to a simulation, which may have properties bearing no relation to the real system. User interfaces fare better in this arena for two reasons. First, there are rarely any timing interactions. Typical user interactions are so slow compared to the processors with which they interact that mimicking the user's timing is trivial, and variations in time delays are not usually relevant. If you simulate a pneumatic system, you may find that the laws of thermodynamics are not so forgiving.

The second reason that user interface simulations work better than other simulations is that the hardware is simpler from a software point of view than is a control system. When an electronics engineer tells me he will provide access to a LED that I can turn on or off, I have a good understanding of what I will get and can simulate that LED easily. If, on the other hand, the electronics engineer says that he is designing a sensor that can read a single byte, the simulation is rarely so simple. Is there a time delay between selecting the sensor and reading it? Does hardware filtering take place? Should I apply filtering in software? Is the scale from 0 to 255 linear? Sometimes, the only way to get the answer is to use the real system. You may be able to get some sample data that was read from a real sensor, but this is not suitable if you are testing a feedback system in which the input varies over time, depending on the system output.

A carefully thought-out simulation of the user interface allows the user, or a script, to generate requests that would be possible from the real user interface. In order to implement the simulation, the lowest-level functions that interact with the user interface hardware are rewritten to read and write to whatever alternative you choose. It could be as simple as using standard input and output if you run the simulation on a desktop computer. Certain keys on the keyboard are interpreted as buttons on the target device. Calls to the output function may cause text output to represent any changes. For example, instead of seeing the transmitting LED light up, you may see the text: `TransmittingLED = ON`.

This text output allows regression testing. You can store the output in a file. If the test is repeated, you can use a text comparison to confirm nothing has changed. If there are differences, you can highlight and examine them to see if they either reflect deliberate changes or represent bugs. You can also use the text output to see repeated actions. If, due to some bug in the code, a LED turns on twice or a certain line draws twice on a GUI, you may not be able to detect this while looking at the interface. However, it will be obvious from the text output. If the LED is turned on and then immediately turned off again, the human eye may not detect it, but it can be established from the text log. This information is valuable when debugging.

Hardware simulations also force the user to draw a strict line between low-level user interface routines and higher level code. If higher level code must run with the simulation and the real system, there are fewer opportunities to cheat by inserting code into the higher level routines that exploits knowledge of the hardware. This discipline benefits the project in the long term, because the higher level routines remain the same even if the hardware changes lower level routines.

8.3.2 Prototypes

Producing an impression of the user interface on a desktop computer is straightforward for many displays. There is a huge proliferation of tools, some for development of desktop applications, others targeted directly at rapidly prototyping front ends of embedded products. Some tools allow the prototype to run across the Internet and gather data on the use of the prototype, providing valuable feedback to the designers (Strassberg, 1994).

How closely the prototype imitates the real system is called the *fidelity* of the prototype. You can improve the physical fidelity by scanning the artwork of the real device into a bitmap and using that as a background for the on-screen prototype. Increasing the functional fidelity is more difficult, as this involves implementing some of the system's requirements on the prototype. However, functional fidelity is more important when gathering user feedback. Be careful not to fall into the trap of increasing the fidelity so that the prototype itself becomes an onerous software project. The investment in a piece of prototyping work can stifle the developer's ability to see alternatives. The developer wants the current version to succeed because of the time and effort invested. Finding an alternative, even if it is better, means throwing some of the work away. Keeping prototypes cheap and nasty reduces this feeling of investment and also reduces the impression that the product is almost ready.

> **The largest scheduling problem faced in user interface software involves making the prototype look like it nearly works. Moving from a prototype to production-quality software invariably takes longer than anticipated.**

Schedule pressure often makes it seem as if figuring out why one of the LEDs sometimes flashes at the wrong rate is more vital than doing yet another demo of the prototype for curious (and troublesome) sample users or marketing representatives. Get these people to step through interactions, and listen carefully to their opinions. Do this as early and often as possible, because bugs in user interface requirements are not due to a misplaced semicolon. They result from the gap between what you can visualize a user doing and the experience itself. Prototyping ensures that fewer requirements changes are made late in the project.

Many managers cringe at the idea of development time spent on elaborate prototypes. One way you can save time and sell the idea to management is to prototype the interface while allowing as much real code as possible run at the back end. You then have a hybrid between a pure prototype and the simulator described earlier. This is particularly appropriate for graphical systems in which the amount of code that drive's the display is high and the low-level graphics primitives can easily be replicated on a desktop computer. Much of the code used to drive a graphical system is independent of the other hardware attached to the system.

Prototypes are often useful for training after the full system is built. They are particularly useful in training for emergency situations. Stanton (1996) describes training for emergencies in nuclear power plants using imitations of the real monitoring system.

Chapter 9

Usability for Embedded Systems

Specifying a user interface is onerous, even before you consider the technical difficulties of implementing the interface in hardware and software. When specifying a product, the user interface is often the most complex part of the customer requirements. One reason for this difficulty is that many other complex engineering issues are considered implementation issues, because the user is aware of them. How much do you need to know about guiding laser beams when you press the play button on your CD player? By definition, the user is aware of everything on the user interface.

Many examples in this chapter step outside the software domain and discuss the mechanical mechanisms used to drive the software. This is because mechanical examples are often easier to explain and to imagine, although they may have parallels in the purely software parts of the design. Also, because the mechanical controls will ultimately interact with the software, software engineers should take an interest in how they operate. Software has to control the interactions, and how well those interactions work depends on nonsoftware factors, such as labeling, button layout, and the user's physical manipulations. In theory, there is no difference between controlling a piece of software with a mouse or with a trackball. However, the lower levels of accuracy with a trackball may lead you to design screens with bigger buttons. The user interface must be considered as a complete system, and many members of a team have to cross the traditional boundaries of their engineering discipline to achieve this.

In some organizations, usability professionals are involved in user interface design. If you conduct extensive field trials, you may need to have at least one team member involved full time. Some organizations have a usability group and a dedicated usability lab — a room equipped with video and audio recording equipment and, sometimes, even a one-way mirror for observing users. Not everyone goes to that extreme. Usability is a bit like motherhood — everyone is in favor of it as a general principle, but not everyone wants to pay the bills. The amount of literature on the costs and benefits of usability and how to justify it to management illustrates the number of companies that do not budget for it. Large organizations that cater to a mass market can afford a dedicated team. Economies of scale are in their favor, and the market size dictates that the one-time cost of usability research is spread over many customers. Smaller organizations with smaller markets must take a more ad hoc approach. The most important problems can be discovered with informal tests and demonstrations. If a change requires a statistically significant study to establish its validity, the improvement is probably only slight. If the improvement is large, it will be immediately obvious from a few users' reactions.

One reason why it is so hard to evaluate the user interface is because it is so difficult to measure. The number of messages that a network router processes in a minute can be defined as a number that marketing can demand and engineering can deliver or negotiate — it is a quantity that all involved can understand. A bigger number is better but usually costs more. Interactions are not so amenable to measurement. Neither the engineers nor the marketing team are able to quantify the usability of a system and say, "Yes, it is 75 percent usable for a novice and 82 percent usable for an experienced user." Not only that, but for a given change, everyone concerned may disagree about whether the change improves usability or diminishes it. To add to the difficulty, some decisions are made before there is a system to test. Even when a system is available, the kind of hard data common in other areas of engineering may not be available, and the disputes continue.

Many engineers do a fine job in usability without explicit knowledge of the principles that their decisions are based upon. Other good engineers fail but cannot define exactly what is wrong with the interface except that customers complain that it is difficult to use. Some interactions seem right, some seem wrong. Sometimes, following examples that work elsewhere keeps a designer on the right path. Pick enough right examples to copy, and you might get a usable interface. This chapter discusses why some of these decisions are right and others are wrong, in hope that you can evaluate your decisions in a more scientific manner.

Embedded systems designers face a greater challenge than software engineers developing for the desktop. On the desktop, the environment has been defined — whether the platform is based on text, Windows, or X-Window. An embedded system may use a piece of hardware for input or output in a way that it has never been used before. Standards for ease of use are also higher. On the desktop, the developer assumes that the user is familiar with the computer. An embedded system is often designed to hide the fact that it contains a computer. A desktop application can be compared to other applications, while an embedded system can be compared to a tool. Try competing with the ease of use of a hammer.

Another reason that usability issues for embedded devices differ from those for the desktop is that relationships between the user and interface are more diverse. On the desktop, you can assume that the user's attention is entirely on the computer. An embedded device may compete for attention with other things, people, or tasks. The user may interact intermittently with the device, and the state of the front panel during the breaks may have to be considered. Sometimes, the user may interact with the device, but not have time to actually look at it. There may be more than one operator, as in the case of a power plant control room or the pilot and copilot in a cockpit. These relationships may not be apparent from the device's technical specification or from a survey of competing devices, so you must observe the device in real use to fully understand the nuances of the interactions. In the context of human–computer interaction, *ethnography* is used to refer to an approach that develops understandings of everyday work practices and technologies. This approach emphasizes observation of the user in the normal work or use environment rather than depending on survey and interview data (Monk, 1995).

There are two fundamental approaches to usability: usability by evaluation and usability by principles. Usability by evaluation involves dissecting a design to find its strong and weak points in order to improve it. Although this is necessary to validate the product, it is not the only way to invest in usability. Historically, much usability work is done this way because it is not considered until someone realizes that the product is hard to use — which may not occur until the product, or a sizable portion of it, is built. At this stage, it is often too late to make the changes necessary for dramatically improving usability, but the lessons learned can often be applied to a later project. The easy part of usability by evaluation is criticizing the current design; the difficult part is deciding what would improve it.

Usability by principles decides ahead of time what usability properties will be desirable on this interface and what types of user will use it. As I describe user properties in the next section and interface properties in the following section, you should recognize many issues from your own customers and products. Along with the principles I present, you will discover many lower level properties specific to your product area. Car dashboard designers may use the phrase *eyes-off-road time* when discussing whether an action can be performed without looking at the dashboard. Other products have names for modes: service mode, fail-safe mode, distance visibility mode. Some phrases transfer across a family of products; some are bound to a particular device. Many of these phrases are meaningless to the user, who may never think about the features in the same way as the designer.

By naming and defining these principles, you are equipped with a language that allows you to discuss and document the product usability features more powerfully. The language also encourages the transfer of usability concepts from one product to another. More importantly, it allows you to decide what you want from the interface before you begin its design. The desirable properties provide a set of parameters to apply to any interaction. Vague principles, such as easy-to-learn, still lead to debate. But the more concrete properties, such as whether the driver must take his eyes off of the road, have solid meaning. After a few attempts at designing part of the interface, some principles may have to be revised if they are unattainable or cause compromises in areas such as performance or safety. In that case, you know what usability you have sacrificed and what you received in return.

9.1 The User

You should construct a picture of your typical user. Some factors, such as typical gender, age group, occupation, and disability, may be self-evident. Psychological factors are more difficult to tease out. You may end up with more than one profile. Some of the information for profiling the user will come from the marketing department. Much can be learned from customer complaints. More than anything else, you should talk to the users themselves to discover their needs and desires. Some Honda engineers even moved in with families to observe how they use the family vehicle.

9.1.1 How Keen is Your User?

Users may desperately want to learn how to use your device, or they may prefer to do without it. In some work situations, the user may be instructed that he must use your product, or the product may be available if the user feels it improves work for him. A user may wish to learn a few basics, such as how to play a video on the VCR, but may have no inclination of how to learn more sophisticated features, such as timers. If the user has the option of not using your product after the first use, as with a photo booth or other pay-per-use devices, then you must focus on the novice user to ensure that he is satisfied with the device from his very first encounter.

McGregor (1985) provides two people-categorization theories: theory X and theory Y. Theory X states that people are fundamentally lazy, do not want to work, and must be coerced into doing the right thing. Theory Y states that people like work, derive satisfaction from doing work well, and strive to improve, so theory X suggests that people probably will not read the manual or heed warnings on the user interface if it involves more work, such as, "Always check that filters are clean before proceeding." Theory X dictates that you give the user few options and automate as much as possible. Why give the user extra information about the process if he will not read it?

Theory Y suggests that adding advanced modes of operation to the device is worthwhile because the user wants to learn them once he has mastered more basic operations. Extra information is good if the user can learn more about the internal workings of the device and, therefore, make more efficient use of the device.

Although theory X and theory Y were originally based on the workplace, similar principles can be applied outside work — setting the timer on a VCR or retrieving money from an automatic teller machine are hardly fun activities. On the other hand, theory X and theory Y do not provide a foundation for reasoning about video games. There, a different set of motivations are in place.

Adler and Winograd (1992) discuss a similar division of user types. They suggest that equipment design in a manufacturing environment may be skills-based or technology-based. Skills-based design sets out to increase productivity by leveraging the operator's skills. The new device should enhance the operator's skills, not replace them. The alternative approach, technology-based design, seeks to automate as much as possible, reducing the operator's skill level accordingly. Adler and Winograd strongly favor the skills-based approach and cite much evidence that it increases productivity.

How do these theories affect the design? Consider a tester for a simple component, such as a resistor. The simplest tester might have a green light and red light to indicate pass and fail, respectively. The user feeds the resistors in and takes them out, placing them in the pass or fail bin, appropriately. The operator requires a minimum of skill. An alternative is to replace the two lights with a needle that shows the resistance of the component. Green and red bands on the needle's background can show whether the resistor has passed or failed. As a technology-based design, this is worse than the original, because the operator may make a mistake reading the needle. He also has to decide if the needle lies on the boundary between red and green. However, as a skills-based design, this is an improvement. The operator has more information. The operator may notice that the last 10 resistors failed but that they all had exactly the same resistance. This may indicate that the resistors came from the wrong batch or that a manufacturing problem had recently appeared. The operator can access information that allows more meaningful monitoring of patterns than the simple pass/fail test.

The skills-based approach is preferred in highly skilled areas, such as cockpit controls. The autopilot can take away from the level of skill of the pilot, but studies indicate that pilot error increases with over-automation. As the pilot has fewer and fewer duties, skills required in an emergency become rusty. Boredom lowers the awareness of unusual airplane behavior and also lowers response time once an emergency is detected. The load must be balanced between the pilot and cockpit electronics to ensure that the pilot is active, but not overloaded.

Although the skills-based approach is advocated by Adler and others, sometimes you should avoid this approach. In a workplace with high staff turnover, minimizing training time may be vital, and skills gained by employees may provide little benefit to the company if those employees leave.

You should not worry about whether theory X or theory Y more accurately reflects the human condition or whether skills-based or technology-based design is fundamentally better. However you should decide which attribute is more appropriate to your target user. Be aware that some assumptions you make may prove self-fulfilling. If you choose theory X and provide the user with minimal decision-making ability, your observations of the user may confirm that he has little interest in using the device or in learning more about it. However, if you choose theory Y and provide the user with advanced modes of operation, your observations of the user may confirm that he wants to learn more about the device and use it more efficiently. You have to consider changes in the users' behavior caused by the way they interact with your device.

9.1.2 Roles

Most devices have more than one type of user. Users may or may not have used similar devices; users may want to use the device for different purposes. A toy may be targeted at the four-to-eight-year age group, but an adult must change the batteries. Although these users may be different people, sometimes they are the same person at different times — the slightly older child who changes the batteries himself may fill both roles. For this reason, I refer to the role the user plays rather than to a category a user belongs to exclusively.

In the following sections, I discuss a few common roles. Your application domain may contain other roles that are distinguished by job title or qualifications. Often, the user's goals define the role. If a user of a global positioning system (GPS) tries to drive to Aunt Mildred's house, his role is different from the expedition leader who uses the same system to navigate a mountain pass. The GPS system may not be able to distinguish between the roles, but the designer must consider them both.

9.1.2.1 Novice and Expert

Some device users are novices; others are experts. Most will fall somewhere in between. You may or may not be able to have a separate mode of operation for experts and novices. Generally, such schemes do not work well because a user does not want to make a decision to transition from one mode to the other. Users may also wish to be expert in some features, but not in others.

Although the novice and expert may not map to complete modes in the user interface, both roles may be considered when performing walkthroughs. Sometimes, you need to consider only one of the roles. Some help messages are needed only by the novice. Some features are requested only by the expert. You should consider the fate of the novice user who accidentally enters an advanced feature. Will he know how to escape?

This often leads to disputes over how much information should be displayed at various times. Too much information baffles the novice; too little information frustrates the expert. You can distinguish the information by color or size of text, or hide the details until the user selects them.

There is no easy way to ensure that each role is given appropriate priority in the decision-making process. Initial field trials involve novices, simply because the device was not available before. The marketing team's desire for more features so that the device can have a longer feature list than the competition tends to favor the expert user. As you consider all input to the user interface design, you must decide which inputs favor one role and which favor the other.

There is a trade-off between power of expression and ease of use. These are areas in which the expert user feels limited if the design is oriented toward the novice. For example, compare the UNIX shell with a windowed interface that has a file manager. The file manager allows you to select files and perform operations on them, such as moving and deleting, by manipulating them with the mouse. The shell command line is more expressive. Regular expressions and wild-carding are used to state exactly which files are to be deleted, and operations can be made recursive throughout a directory tree. However, the command line is more difficult to use. Novice users prefer the graphical file manager approach; the experienced users prefer the power of the command line.

Command lines are unusual in the embedded world, but similar trade-offs apply. Consider a TV remote with up and down buttons for the channels. They are easy to use but not powerful enough to jump straight to the desired channel. If you have a numeric keypad, the user can type in any channel. But the user must learn how to manage two- or three-digit numbers. Does 5 followed by 3 mean go to channel 5, then go to channel 3, or does it mean go to channel 53? The interaction is more complex but more powerful. In the case of the TV remote, it may be trivial to provide both methods; at other times, you have to choose.

9.1.2.2 Are You Being Served?

As customer expectations for timely service and repair increase, you should think about the hard-pressed service team. They know the device better than anyone, so they will appreciate some cleverly concealed features that give them extra access to the internal state of the device. In some cases, the device has a service mode in which only features designed for the service technician are available. You may want to supply the service technicians with an overlay that sits on top of the controls and relabels them to read the diagnostic duties available in service mode.

Even if you do not include a separate mode for the service technician, you may want to make the system robust enough to run with the lid off and with some cables disconnected. Disconnecting a cable and checking the voltages can provide useful information if the system still drives the signals on that cable. If the system is designed to shut-down when it detects the lid open or a cable disconnected, perhaps a jumper or dip switch could override it. However, if it is overridden, it should be obvious from the user interface so that it does not remain in that state after the service is completed.

9.1.2.3 Privileged Users

The concept of superusers on UNIX and other system administration roles have long been a facet of desktop computers. Cash registers may need keys that turn a lock to allow a teller to start his shift. On some devices, a combination of buttons must be pressed to allow access to a normally hidden feature, which may take the form of extra options on a text menu. The options may allow the device's language to be changed or the time of day to be updated. Be careful that these options are removed in some sensible way. You do not want the privileged functions to remain available after the privileged user has walked away. Another way to protect functions is to conceal their controls behind a panel that must be removed with a screw driver or by opening a lock.

In some cases, especially when money is involved, great care should to be taken to protect employers from theft and employees from accusations of theft. In other cases, you will want to add a layer of protection against certain facilities to avoid accidental changes during routine use.

9.1.3 The User's Conceptual Model

As a user becomes familiar with a device, he builds up an impression of how he believes the device operates. This image is a conceptual model, sometimes called a mental model, of the device. The user does not need to understand how the software or electronics works. The model need not be an accurate representation of the actual operation. What matters is that the user's model of events allows him to decide how to control the device. The model needs only to describe the device at the level that can be manipulated at the user interface.

When a user uses an ordinary tape deck, he knows that he has a finite amount of tape and that the motors turning the tape have three speeds: play, fast forward, and rewind. This much detail allows the user to understand why he cannot jump to a random spot on the tape instantaneously, as is possible with a compact disk. When the user buys a tape deck with auto-reverse, he understands that the tape can play the second side after the first side is completed. However, it is not feasible for the device to immediately replay the side that just completed. The tape deck designers want to reinforce this model, so they make part of each reel visible through a clear slot on the tape and a matching slot on the door of the tape deck. The user can estimate how much time is left on the tape by looking at the amount of tape on each reel. On the other hand, the movement of the heads to touch the tape when the play button is pressed is a detail that does not add anything useful to the user's conceptual model. The user may know about it, but the information does not make the device easier to use. Therefore, the movement of the heads is usually not visible from outside of the tape deck.

By deciding beforehand which parts of the design will be part of the user's conceptual model, you can make decisions about the terminology that can be used, and what diagrams are appropriate on the device or in the documentation. A conceptual model is reinforced each time a device displays another attribute that matches that model. A conceptual model is reinforced when the user makes a prediction about the device based on the model, and that prediction is later proven correct. The user loses confidence in the model if an attribute of the device contradicts the model. This may lead the user to form a new model that accommodates the new information. You as the interface designer should know which model works well for the user and reinforce that model in the documentation and the interactions at every possible opportunity.

In embedded systems, a large part of the model may be based on a mechanical system, so it is easily represented in a diagram. Parts of the system that are under software control may be more difficult to represent.

As a user becomes familiar with a device, he builds a set of rules. For example: "If I am in mode A, then I can transition to mode B, but not to C". The implementor may use an FSM to define this rule, along with several others. A diagram of the FSM in the user documentation may be helpful, but many rules are learned through trial and error. The user also forms a model of the device's storage. Some parameters are remembered through a power cycle and some are not. The user develops a model of permanent and transient memory, even if he has no concept of RAM, nonvolatile RAM, or flash memory.

One product I had on trial received several complaints about the accuracy of the volume measurements. They were well within the specification, and I thought that users should be used to seeing volumes of 1,500 milliliters vary by a few milliliters on the competitors' machines. I knew that the competitors machines could not have been that much more accurate than ours. When I looked at one competitor, I found that the product avoided the problem completely by displaying volumes in liters to one decimal place, rather than milliliters. Therefore, minor deviations were not visible. The user's conceptual model did not include mechanical tolerances. This was not raised as an issue earlier because all the people dealing with the machine had engineering backgrounds and accepted small percentages of inaccuracy as intrinsic to any mechanical system.

You can easily violate the user's conceptual model with text messages. For example, a product with two CPUs displayed the message, "Test mode requested." This was displayed just after one CPU requested the other to enter test mode. There was usually a short delay, so it was important to let the user know that he is expected to wait. At first, I could not figure out why the message did not seem right. The problem was that the request for test mode was sent from one CPU to the other. The user did not need to know that the device contained two processors, nor could he be expected to appreciate that one processor had to request test mode so that they could synchronize. "Entering test mode" would have been more appropriate. As it turned out, the user usually thought that it was confirming that the user, not one of the CPUs, had requested test mode, so very little real confusion was caused.

> **Never lose sight of the user's conceptual model; it may be very different from the engineer's model.**

9.1.4 Observing Users

One important part of designing a user interface is observing users and interpreting their comments. Surveys, interviews, and observations of users interacting with products or potential products will help you to get inside the head of the user. Most literature about interface design advocates the inclusion of users at all stages of the development cycle. This approach is necessary if the interface is the central component of the product, such as a flight simulator or a virtual reality application. The need might not be so great for applications with less demanding interfaces. At the start of your project, make sure that you are exposed to the needs of the user and learn how to look at the device from the user's point of view. This outlook benefits you throughout the project.

9.1.4.1 Users, Users Everywhere

Before you can observe users, you must find them. An adage in the psychology world is that most published psychology is the psychology of college freshmen — because the most convenient and plentiful supply of guinea pigs for academic research is among the undergraduates in the researchers' own institutions. If you get the engineer who works at the next bench to try out your VCR interface, the complexity of programming timers may not be a challenge. But a broader group that contains nontechnical people might show that there is greater difficulty in this area. To avoid next-bench syndrome, spread your net wider. Eastman Kodak sent observers to Disneyland to take pictures showing how camera owners carried and used their cameras. For office products, typical users may not be so far away: Xerox has been known to mount video cameras above photocopiers in its offices. Tests like this, in which the user does not realize he is being tested, avoid the problem of users changing their behavior when they are being tested. The tendency of the subject to work harder when they know they are being observed is known as the *Hawthorne effect*.

You can easily find sample users for mass market devices, such as cameras and photocopiers. If you require feedback from pilots, doctors, architects, or other specialized users, you may need to form special relationships with a small group of your customers. The danger of such groups is that the group size is not large enough to yield statistically significant results. Also, the customers you consult may not be typical, because companies often seek out the leaders in the field to consult on market needs. These users are often so familiar with the type of device being studied that their use does not reflect the typical user who has less experience. These domain experts' needs are skewed toward more powerful features that may be relevant to only a small percentage of the user population.

Another danger is in having the marketing group talk to the managers buying the system, putting two degrees of separation between the real users and the engineers. Figure 9.1 shows a bad, but not uncommon, situation. The engineers must bully their way to the end users to find out the real needs. Managers often try to convince themselves that their employees are unskilled and easily replaced. When you talk to workers about the job they do, many skills emerge that receive no formal recognition and are not documented in the company's procedures. In a nonworkplace situation, you may find that the product distributor is the immediate paying customer for your company, and his perceptions of the end-user may also be flawed.

Users often try to cheat the interface. In medical ventilation systems, users may lie about the volume of gas that is to be delivered in order to compensate for a leak in the system. If they estimate that they are losing 250 milliliters, they may dial in 1,250 milliliters, expecting to get 1,000 milliliters delivered to the patient. This suggests that an automatic way of compensating for the leak is beneficial. You find out about these only by talking to the end-user, not their manager or the purchasing department in the hospital.

9.1.4.2 Getting Feedback

To conduct an experiment scientifically, all users should be presented with the device, possibly some user documentation, and a task to complete. Videotape the users' progress for later analysis of the users' mistakes and the timing of parts of the task they were given. The user may be asked to think out loud so that he can give further insight into why he performed certain actions. Written questionnaires or an interview provide more feedback afterward. Wiklund (1994) documents many such scientifically conducted experiments. If the interfaces of several competitive products are compared, the user might not be told the name of the company conducting the experiment so that it does not affect his responses. The closer the environment is to a real

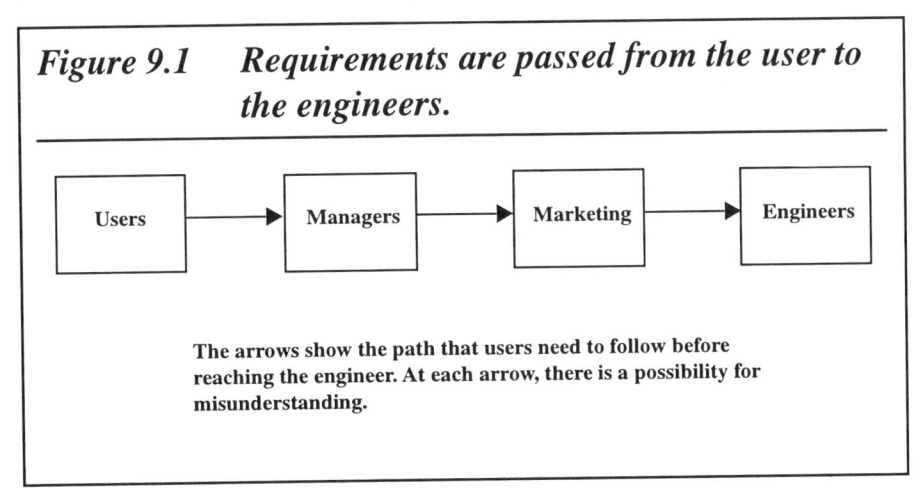

Figure 9.1   Requirements are passed from the user to the engineers.

Users → Managers → Marketing → Engineers

The arrows show the path that users need to follow before reaching the engineer. At each arrow, there is a possibility for misunderstanding.

situation, the better. Automatic teller machines in very cold climates need bigger buttons so that they can be used with gloves on — something that could not be discovered with users in a heated observation room.

Such scientific experiments can be costly, and it is hard to make them meaningful unless you have a working product for the user to operate. Sometimes, a demonstration followed by an audience discussion is more feasible, especially at the early stages. These demonstrations are often presented by an engineer who is responsible for developing a significant part of the interface. A user may be less critical of the interface in the presence of someone who may feel personally criticized. Similarly, a developer who becomes defensive during such discussions can hamper proceedings. It is often better when the designer is not present or does not play a central role in the discussions. If the user does not know the team, the designer can say, "This is what the folks in engineering came up with, but I am not too sure about it." The demonstrator is distanced from the design and the audience feels less threatened. By being mildly critical of the interface, the presenter sets a tone that encourages others to express what they feel, without fear of offending. You may wish to point out to the user that the interface is being tested, not the user. This avoids the situation in which the user pretends to understand something, to avoid appearing foolish, when it is exactly that failing of the interface that you want to identify.

In the discussions that follow a trial or a demonstration, you should have a strict agenda. When you ask users what they think of an interaction, they tend to say that they did not like X and then they describe the interaction that would work. You should find out why X was not liked. Do not let the user spend too much time describing the solution. The group may move to a discussion of the user's hypothetical interface rather than the one being demonstrated. The user's solution may be valid and, if so, the design team can discuss it later. Demonstrations are not good opportunities for gathering solutions because an example activity is often chosen for the demonstration. The user describes an interface that suits the particular example, but it may not represent a good general solution.

Try to find out two things from the discussion: the needs of the user and the faults in the current interface. Let users talk about whatever seems to them a good idea. However, do not allow these ideas to dominate the exercise.

Find out the user's needs. Do not ask him how he would like to interact with the device, but what problem he is trying to solve.

9.1.4.3 Paper Prototypes and Working Prototypes

The presentation to the sample users may consist of a storyboard display in which you show the user a picture sequence of the proposed interface. However, paper prototypes suffer from several drawbacks. A storyboard always gives the impression of a slow interface. When a different picture is shown for each step, it may be hard for the user to discern which parts of the picture have changed, and he is forced to scan the whole display each time.

After being shown a four-key sequence, the user often responds, "You mean it takes me that long just to initialize the system?" The user's requests are based on reducing the number of steps rather than making the interface more intuitive. If the user were to press the four keys on a working prototype, it would take only a few seconds and not seem as laborious as turning several pages.

With a functioning prototype, the users are looking at one display and any changes are noticeable. The human eye easily detects movement against a static background. Using transparent pages over a background may help to emphasize the changes on a paper prototype, though it is still not completely satisfactory.

Unfortunately, paper prototypes emphasize appearance. The user can hold up one picture and analyze it in isolation. This leads to interfaces in which appearance is given priority over functionality. This does not meet the engineer's needs, because the functionality must be agreed on in order to nail down the requirements. You can finalize the appearance later, once the functionality has stabilized.

Several tools allow rapid prototyping of interfaces by presenting a graphical version of the system on a desktop computer. Some or all of the functionality of the final system can be built into these prototypes, making the experience realistic for the user. If the real functionality is too difficult to implement, then a sequence can be programmed into the system. If the user follows that predefined path, the interface appears to work correctly. You can also connect the prototype to another computer system, which controls the prototype's output. The second system may be controlled during the demonstration by an engineer who feeds in appropriate feedback to each user action. This is called the *Wizard of Oz* method. See Chapter 8 for more discussion on implementing prototypes.

9.1.4.4 Interpreting the Feedback

Is the customer always right? I may be committing marketing heresy by suggesting that the customer or the user may sometimes be wrong. You should ask yourself the same question many times during preference trials. When a customer asks for a feature, you as engineer should point out the trade-offs. If a display shows a list of records in chronological order, the user may give reasons why an alphabetical ordering is preferable. You know that there will be a performance penalty if the list is sorted each time it is displayed. Users may barely be satisfied with current speeds, and the performance penalty may cause them to reverse the change after they see the next interface revision. Often, you can see these issues ahead of time; a user cannot be expected to foresee the trade-offs.

If you constantly change the set of sample users, your user's may always be at the novice level. At the other extreme, you may always use the same group. The danger here is that they have seen the design in many iterations and are familiar with the history of the interface. That history can sometimes justify idiosyncrasies that a fresh user would find less acceptable.

When users contradict each other, you should establish the reasons. Does one group conform to theory X while the other conforms to theory Y? Do the opinions come from genuine users or from parties representing the users? If the users' opinions diverge, perhaps your approach should be that you cannot make all users happy all the time. If you still must make a decision about which way the interface should work, try to find a similar situation elsewhere in the interface and conform to it, so at least you maintain some consistency, even if the interaction is not to everyone's liking.

Users may have novel ways of measuring whether an interaction meets their requirements. I worked with one medical practitioner who held his breath while he tried to interpret an alarm screen. He figured that if the patient is not breathing while he is trying to solve the problem, then the length of time a person could hold his breath was the target to beat. If he took another breath before he figured out the problem, the information on the interface had to be clarified.

9.1.4.5 Usefulness and Usability

Users may be required to establish usefulness, as well as usability. Do not confuse these two attributes. Landauer (1995) goes to great trouble to point out that making a product usable does not necessarily mean that anyone has productive use for it. Finding a basic use for the device should be one of the first steps in deciding what to build. As the product matures, other uses may emerge or some original ones may be ruled out. This is another reason why you should establish the user's needs and not just the set of interactions he is comfortable with. There is no point in perfecting the user interface of a calculator that works with base-17 arithmetic, only to discover afterwards that no one actually wants one. Technology-driven companies often assume that because a product is technologically impressive, someone will buy it. Not so. In many cases, there are already competitors on the market, so the device's utility, or at least its sales potential, is already established. You should still assess uses for the devices and whether there are uses that your competitors have not discovered.

9.2 The Interface

Now that you know who the user is, I will discuss the interface itself. Most properties described in this section are desirable, and you should try to achieve them. In Section 9.2.8 "Paths Through the Interface" on page 254, you will see some properties for which choices may depend on user characteristics I have already defined.

9.2.1 Robustness

Robustness is often thought of as the mechanical property of an object tolerating rough use and carelessness. A robust user interface is not necessarily physically strong (though that has other obvious advantages), but it tolerates improper inputs or prevents them. A robust interface not only protects the device from accidental damage due to an incorrect input, it also protects the user and the entities the device acts upon.

In 1982, the Sinclair ZX Spectrum home computer's user manual stated: "Nothing typed at the keyboard can damage this computer." This showed good insight on the writer's part. The company realized that it could reduce users' anxiety by reassuring them that they could not destroy their purchase, no matter how many silly mistakes they made. It was also a sign of good computer design. It was robust. Many programmers wrote their first programs on that system, without worrying about the damage a buggy program might cause.

In the desktop world, when you request that a file be deleted and the system deletes the file, you expect nothing else. But what if the initial request were a mistake. A robust interface should not easily allow such an accident. The system should prompts users, "Do you really want to delete *MyLifesWork.txt*?" This makes the system more robust, but it is now less usable because it takes more key strokes to delete a file. An Undelete command provides a better approach. Now the system is more robust because the file cannot be lost easily, and the ease of use is not compromised.

In general, requesting confirmation of an action is a clumsy way to add robustness. Using an embedded interface should be as natural as using a tool from your toolbox. Your hammer does not ask you if you want to hit something just before impact. If it could, I do not think that many people would consider it a more usable tool. Confirming actions becomes so automatic that it adds little protection anyway.

Similarly, error messages or beeps may seem like appropriate responses to invalid actions. If you ask an automatic teller machine for too much money, it displays a polite message telling you that you are not rich enough. The system is protecting the user. Sometimes, you must put limits on the user to prevent him from setting illegal values. These limits protect the device and the user from user errors. Sometimes, a mechanical system has intrinsic limits that cannot be violated — the top speed of a conveyer belt motor may be limited by its circuit. However, an internal mechanical or electrical limit may lead the user to believe that the belt is moving at the entered speed instead of at the limiting speed. By limiting the user at the time of input, you can give more accurate and meaningful feedback.

You should be careful how you word error messages or warnings. Displaying a message that says "Illegal action" not only gives the user very little information, it also suggests that he is a criminal! With some extra effort on the developer's part, you can provide context-sensitive messages that tell what is wrong, such as, "Can't record: no tape inserted."

Now that you have established a protocol for telling the user what he has done wrong, you have improved the product — but the interface is less user friendly. Devices that tell you what you can and cannot do are unpleasant to use. However, you can often get the best of both worlds. On a TV set, a user can press an up or a down button to change channels. What happens when he reaches the highest channel? You can beep and flash if the user tries to go any higher, and he may realize his mistake, but it is more satisfactory to wrap around to the lowest channel and avoid the error message.

Think of the error message as an airbag — it minimizes damage once the accident occurs. You are better off supplying the user with an antilock braking system that might avoid the accident. If a device takes numerical input from a keypad, it may have to reject an illegal input. If you replace the numeric keypad with a dial, the dial can limit the range at its mechanical limits and avoid the possibility of entering an illegal value.

In the last two examples, the system can detect that the input is not normal and takes alternative action. What about when the user performs an action that is a valid input to the system and then realizes that it was not the appropriate action? The Undo command is a popular feature on many desktop applications. However, Undo forces the user to figure out how much will get undone. Will a second Undo go back further into history, or will it redo the undone command? You cannot always afford to hit the Undo button just to find out what will happen.

Sometimes, you can provide a set of actions that allow the reversal of any action. If you move a robot arm to the left, then make sure you can move it to the right just as easily. The effort to undo the action is equal to the effort to perform the action in the first place — leading to a robust system. A tape deck (and they do exist) with a fast-forward button but no reverse is frustrating because the user must turn over the tape to undo the action taken when he wound forward too far. The effort to undo the action is greater than the effort that causes the action — leading to a less robust system. Similar situations arise with mode changes. If you require 10 steps to get from mode A to mode B but only one key press to get from mode B to mode A, then your users will be displeased if they enter mode A by accident. One accidental key press takes 10 actions to rectify. The principle that actions and their opposites should involve similar effort is what Thimbleby (1990) calls *commensurate effort*.

There is one exception to this rule. If mode B is a safer mode than mode A, you may wish to make it difficult to get into the high-risk mode but easy to escape from it. This does not make the system easier to use, but it does make it safer.

In an embedded system, reversing some commands is impossible when the physical action taken by the device cannot be undone. You cannot unlaunch a rocket! In these cases, some confirmation is required. When you implement a confirmation, you must balance ease of use and certainty of intent. If you make the user type in a complex sequence every time he wishes to perform an irreversible action, he is unlikely to do it by accident. However, ease of use is reduced. If such actions are rare and the cost of an accident is high, then the confirmation procedure may be elaborate, such as making two operators simultaneously press a button reserved for this purpose. The buttons can be placed far enough apart so that one user cannot press both. You should avoid forcing the users to confirm so many details so many times per day that they confirm their actions automatically, without actually reconsidering the consequences.

9.2.2 Consistency

Overall design consistency allows the user to learn general rules that can be applied to the whole interface. Interface consistency is reflected in the software and vice versa. Therefore, the code is a good place to detect inconsistency: "I can use this `blinkLight()` function everywhere except this one place. Why? There may be something different about what the user sees, as well."

Consistency lets the user develop general rules about how the interface works. When the user explores a part of the interface he has not previously used, he will depend on these rules. Do not lower your levels of consistency for the more advanced features, assuming that only expert users will use those features and that they will figure it out, regardless. Consistency is of greatest value in these advanced features. The user may rarely visit the more advanced features, and he will not want to learn them from scratch each time.

9.2.3 Affordance

The property of affordance indicates how well a device's appearance reflects what it does. If I hand you a pair of scissors, you notice that one end has proportions that fit comfortably in the hand. The other end has sharp edges, so you are unlikely to hold it there. However, the sharp edges rubbing each other indicate that it is meant to cut something. Scissors afford holding and cutting.

Some things are obviously easy to use because you can see all the controls. A tape deck has a door that opens to reveal two spindles a tape can be placed over. Often, the spindles are visible through a glass panel in the door and are seen even if the observer does not open the door. The tape deck has the affordance of something that can hold a tape. Until you examine the controls, you do not know if this device is for copying or rewinding tapes, or whether it can actually play the tape. If speakers are visible, that is your clue that it actually plays tapes. The buttons that control the tape may be marked Play, Rewind, Fast-Forward, and Stop, giving further information. The observer now knows that the device has three functions and a way of halting the functions.

Conditioning teaches us that buttons are pushed and that dials are turned. But what if the device has a less conventional control? A joystick shows that it can be grabbed by having grooves on the shaft to fit fingers. If a user approaches a device controlled by a touch screen, he may not immediately realize that the screen's surface is touch-sensitive. Making the on-screen buttons three dimensional indicates to the user that they can be pressed down, just like a mechanical button.

If the front panel contains buttons and dials, the user knows that the buttons can be pushed and the dials can be turned. But what do those buttons actually do? Labeling is a delicate art. Sometimes, a button performs more than one action and requires two labels, such as Stop/Go. The double label looks awkward. Power does a good job of replacing On/Off, but such replacements are not always available. If a dial is graduated, then naming the units may replace the need to name the function. A dial graduated in degrees does not need to be marked Furnace Temperature (assuming that there are no other temperatures settable on the furnace).

One common mistake in industrial design is placing buttons in a regular pattern to make the appearance of the device more symmetrical. Designers line buttons up like soldiers, each one resembling the next. This makes sense if each has a similar meaning, such as each one representing a TV channel. If the buttons perform separate functions, group them according to function. Use bigger buttons for the more popular functions, and keep the rare-but-nasty functions out of the way. If buttons are usually pressed in sequence, arrange them in left-to-right order. The user reads this way, and it is the way he scans a screen or display. Arrows, as seen in Figure 9.2, are even more useful if the path you want the user to follow is not as natural as left to right.

9.2.4 Icons

You can use icons to avoid translation issues. Icons often occupy less space on the front panel than the corresponding text. Unfortunately, unless an icon is already established, you may find it difficult to introduce one. Even when icons are standardized, many users do not know what they mean. Do you understand the icons used to label

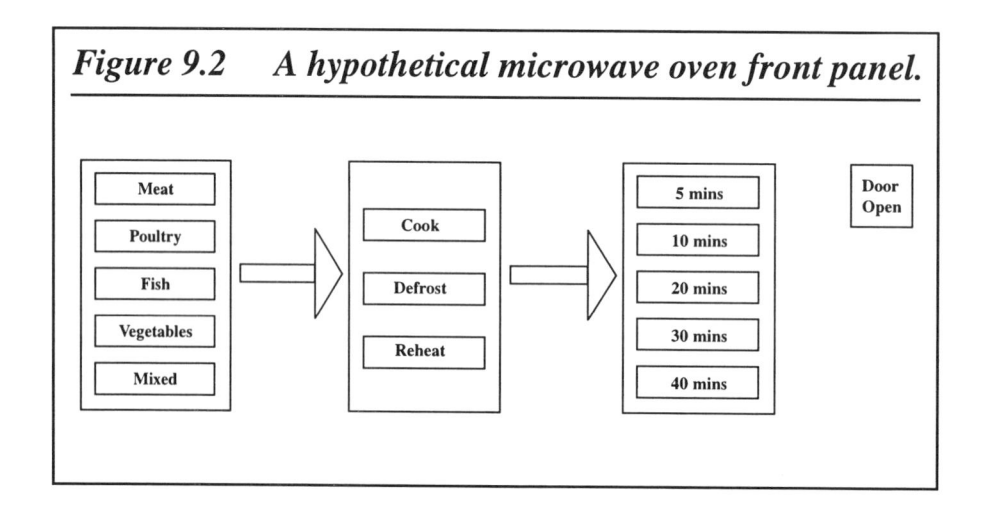

Figure 9.2   A hypothetical microwave oven front panel.

garments safe for ironing, spin drying, and so on? Probably not. How many traffic signs do you understand if they are shown out of context? Most drivers would do well, but try showing them a set of traffic signs used in a different country. Again, the failure rate, without training, will be high. You can expect to reduce the usability of the product significantly, especially for first-time users, if you must introduce your own icons. Video recorders have forced some symbols on us for play, rewind, and so on. They are so widely accepted that they are often imitated on desktop computer applications, instead of text. You may find that the icons you design are not as warmly welcomed. Icons become bearable only after frequent use.

Icons are widely used in desktop applications, such as word processors. On desktop computers, many functions available via buttons that contain icons are also available from pull-down menus, so the user need not understand all icons immediately. GUIs can also provide *bubble help* as the mouse passes over an icon. This allows the user to read the associated text without sacrificing the real estate a permanent display of the text would occupy. On a mechanical button, those luxuries are usually absent, so if your market contains infrequent users or if user confusion has a high cost, such as in safety-critical systems, then avoid icons on front panel labels. If you must use icons, provide on-line help or a cheat-sheet that shows the icons and gives a quick explanation. A cheat-sheet can be attached to the device, or it can be printed on manual covers. The documentation should also state what the icon represents.

The picture is often so small that even after you learn the function of the key, you cannot associate the picture with the action. Does Figure 9.3 show a pot of paint pouring or a mortarboard worn on graduation day?

When the user knows what is in the picture, he generally knows a name for the thing or a noun. The verb of the action statement is harder to capture in a picture. An icon showing a battery could mean the battery is recharging, the device is running on battery, or the user should check whether a battery is present.

Icons are better than text at indicating the user's physical actions. A picture of a card being inserted into the slot on an automatic teller machine clearly indicates where the card goes in. Otherwise, the user can mistake it for the slot that delivers the

Figure 9.3   Paint pot or mortarboard? You decide.

money. This picture works because there is enough room to display a convincing picture. Also, a picture is the most effective means of showing the card's orientation and direction as it is inserted. These symbols work because a user can imagine what a person doing the same thing looks like. A graphics screen might show a picture of a user opening a door on the photocopier when some service is required, or such a picture on the front panel can be illuminated at the appropriate time.

9.2.5 Surface Area

As the feature count in a device grows, the design is likely to hide many features. This is often driven by industrial designers, who want a simple form, and mechanical engineers, who want fewer parts. So many of the less frequently used functions end up under a menu or reusing buttons that are already available. Consider a telephone that can access voice mail. If you can only use the buttons with digits on them, you have no clue that the device is capable of accessing voice mail. Do not think that you are doing the user a favor by making it look like any other telephone that he uses. The phone would be far easier to use if the buttons Enter Voice Mail, Next Message, and Delete Message were added. The phone may look more complex, but users not interested in the voice mail options simply ignore the extra buttons.

The front panel of a device consists of several controls and displays. A device with more dials, buttons, and displays has a greater surface area. The more user interface functionality that is visible to the user, the easier it is to learn the whole device and the more obvious it is what the device can do. The idea that users will be scared of a device with too many controls is a myth. Most people drive cars with dozens of controls. Users resent many controls if they lack organization and if the controls do not provide adequate indication of their purpose. The worst designs result from taking a product that has a fundamentally complex interface and trying to deliver it through a simple front panel. Devices are difficult to use because they have too few buttons, not because they have too many. If your software must interpret the same button in many ways, depending on the context, you may be cramming too much control into one component. I call this *multiplexing* a control. It is harder for the user if he has to decide, based on the current context, whether button 2 means "exit voice mail" or "delete message."

You can also multiplex output. A LED can mean "power-on" most of the time, but it can flash to indicate "low battery," at other times. One numeric display can show temperature at one stage and the time of day at another. You should give the user a hint, such as an A.M./P.M. indicator, to show which mode the device is in. A second display increases the surface area, making it a better match for the functionality. Twice as many displays makes the user's life easier, not harder, assuming that the displays are properly labeled.

9.2.6 Compatibility

An interface has three levels of compatibility: compatibility between what the user expects and what the user gets; compatibility between different products of the same type; and compatibility between the device and its surroundings, and the devices with which it has to cooperate.

A lever-operated press moves down when the arm is raised and up when the arm is lowered. It works this way because of the levering mechanism. This is good engineering, but bad usability. Although the user may learn that the arm moves in the opposite direction to the press, in an emergency he may revert to the more natural mapping and cause an accident. Making the actual behavior compatible with the expected behavior can be easier in software in which the engineering issues can be hidden. So Up buttons should be above Down buttons, and Left buttons should be to the left of Right buttons. Similarly, high and low limits should be displayed in the appropriate vertical alignment. If a set of controls are near each other or have similar coloring, the user expects them to be related. In general, you should not surprise the user.

Compatibility between products is not so simple. The history of products in your market may dictate that certain practices are followed long after a better way has been found. The QWERTY layout of most keyboards is an example of this. Sometimes, there are arbitrary differences in the marketplace and you must choose which trend to follow.

Standards groups, so active for the desktop, have made only minor inroads in the embedded world. Some attempts have been made at standardizing alarms sounds and alarm lights in hospitals, but that is a long way from standardizing the whole interface. ISO 9995 sets a standard for the arrangement of letters on telephone buttons and other numeric keypads. The symbols on a VCR's main buttons (see Figure 9.4) have also been standardized, though the rest of the VCR's controls can vary even within one product range.

Some standards are not so quickly accepted. ISO 8601 is a standard for date and time formats. It dictates a yyyy-mm-dd format for dates. However, Europe currently uses the dd/mm/yy format and the United States uses the mm/dd/yy format. The ISO standard is a more logical format because it starts with the most significant unit and

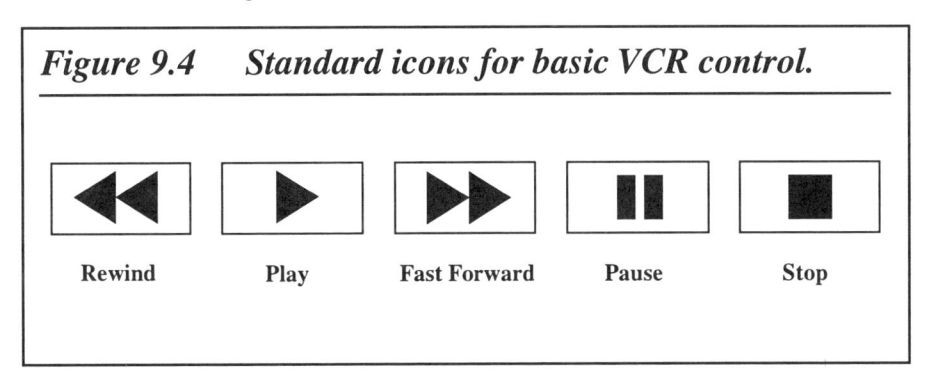

Figure 9.4 **Standard icons for basic VCR control.**

Rewind   Play   Fast Forward   Pause   Stop

moves to successively smaller units as you read from left to right. The standard corresponds to normal numbers and telephone numbers (country code followed by area code followed by phone number). It also has the advantage that dates can be sorted chronologically simply by sorting them alphabetically because the most significant unit (year) is to the left. However, I have not seen any programs or equipment use this standard — mainly because the other formats are well established and impossible to dislodge.

If you have no standard to keep the behavior of competing products in line, then at least try to ensure that all products in a family or from the same company are compatible. Compatible mental models and behavior is more important than compatible appearance. It is more important that the user be able to transfer from one product to another easily by matching interactions rather than matching the color or company logo.

Compatibility with older products sometimes stands in the way of a consistent orthogonal interface. Competitors or predecessors of a piece of equipment may set a precedent for the way some operations are performed. For some common operations, you may have to follow the established norm, even if that does not fit in with the way other operations on your interface behave. Engineers often find this frustrating — their elegant design is soiled by what is seen as an artificial and unfair requirement created by history.

When the interaction must change you may find it advantageous to change the physical appearance deliberately so that users do not expect the same behavior. On an older product I had one red light and one yellow light, which had the words *Alarm* and *Caution* written on them. For the newer product, the rules governing when the lights turn on and when they flash changed because I had to comply with a new standard handed down by a regulatory body. However, the front panel design was kept consistent with the old product by keeping the words *Alarm* and *Caution*. This causes confusion because users, who had used the older product for years, expected the same behavior when they saw the alarm names. If the names had changed, they would have gotten a clue to expect different behavior.

A product must also be compatible with devices that might be used next to it. If you sell components for a stack stereo system, the on/off button for each one should have a similar positioning. If you sell TVs and VCRs, the buttons for changing channel should work similarly.

You must also try to be compatible with the surrounding environment. If the device is used in a noisy environment, then quiet alarms are not appropriate. On the other hand, a pocket calculator might be used by students in a quiet library, so you want to avoid loud key click noises. Sometimes, you cannot guess the environment in advance. One automatic teller machine lobby that I have used has mirrored walls to prevent customers from feeling claustrophobic. Unfortunately, the same mirrored surface makes the user's key presses, including his identification number, visible from almost anywhere in the lobby!

9.2.7 Setting the Pace

Many people are afraid of computers and computer-controlled equipment. They are afraid that the device will eat their credit card, lose the letter they are typing, or tell them they are stupid. It seems absurd that people get offended by their interactions with an inanimate device, but they do. The designers must reduce this apprehension.

If the interface sets the pace of the dialog, it gives the impression that the computer is in charge. Consider an automated answering service that gives the caller a number of options: "Dial 1 for sales inquiries; dial 2 for technical support; dial 3 for reception ...." These services are frustrating because they take so long. The listener can understand several concepts in a few seconds. But because the medium is speech, the information takes many seconds to communicate. Sometimes, if the desired choice is not on the list, you must wait for silence at the end of the list to let you know that there are no more options. If you pick a wrong option, you often must wait until the end of the list to find out how to go back a step (if it is possible at all). The caller's frustration at the slow pace is compounded by the fact that he is paying by the minute for the call.

Long delays while the device processes an event also annoys the user. He may not be sure whether the button was pressed properly. So he may press it repeatedly, hoping for a response. One approach is to give an acknowledgment of the key press immediately and present the final results later. Schneiderman (1993) describes a telemedicine system in which a microscope is remotely controlled, using direct manipulation of a graphical interface. Movements often led to overshoots because of the delay between an instruction and feedback to the operator. One improvement changed the move-instruction format from direction and distance to absolute coordinates. Another improvement made the queue of requested actions visible to the user, so he knows the commands had been accepted and were pending. This reduced anxiety about whether the system had stopped functioning or whether a command had been lost.

At the other extreme, if a device with a text interface responds immediately to a large task, the user may be thrown off balance. One reason is that the user's mental model may indicate that the device has a lot of work to do, such as check the user's password with a central computer. If the password is buffered, the device may respond immediately, leading the user to become concerned whether the device is working correctly. Also, if responses are immediate, the user feels that he is being interrogated. As soon as he enters one response, he is asked for the next. In these cases, you may wish to add a delay to slow the interface a little, making the user more comfortable. If you choose this option, be sure to check how the delay affects throughput for the expert user.

9.2.8 Paths Through the Interface

As the user navigates from one point in the interface to another, he follows a path. Like following a trail, the way may be easy or hard, short or long. The following sections discuss properties of interfaces that dictate the types of paths that are available to the user.

9.2.8.1 Directed Interfaces

Some interfaces strongly suggest a direction. A question-and-answer session provides an interaction in which by the user interface dictates the direction. The user may have the option of breaking out of the sequence, but the questions themselves suggest that the next appropriate action is to provide an answer. This type of interaction is considered *directed*. A car dashboard is a less directed interface. You start the ignition, and there are a number of things you could do. The sequence of putting it into first gear and pulling out of the parking space into traffic involves lots of options — you could have gone into reverse, the lights may have been turned on, the wipers may have been selected. The interface does not suggest any path more than any other. This is a *nondirected* interface.

A more directed interface suits a device with one simple goal. It also suits a novice, who may not want to make many decisions. A nondirected interface provides more power to the user who knows how to navigate the device's features. He can go directly to the control he wants without following a predefined sequence. Sometimes, an interface changes from one to the other. An automatic teller machine gives you few or no options until you enter your card and identity number. This is as it should be. The machine should not provide any service to a user who attempts to by-pass this step. Once the card is validated, the interface relaxes and allows the user more flexibility. The interface is directed enough so that the novice does not get lost, but several options are available at each step.

9.2.8.2 Multithreading

Some interfaces allow many independent paths through parts of the interface to be active simultaneously. Consider a hi-fi system with a CD player, tape deck, tuner, and amplifier. The tape deck can rewind while the user changes channels on the tuner. The amplifier may take input from the CD player, so that is the audible source of music. Each piece of equipment is allowed an independent thread of control. This is analogous to multithreaded software. In most hi-fi systems, you might consider the ability to multiplex an accident of the design of independent components incapable of communicating with each other. A smarter system might direct the amplifier to the last device used and disable all others. This avoids accidentally leaving a tape playing while you are listening to the radio. However, you lose the ability to rewind a tape while listening to the radio, something users may find useful. The conclusion from this investigation of the multithreaded hi-fi: Multithreading makes using the device more challenging, but it offers more power to the user.

Multithreading an interface does not necessarily require multithreaded software. The state of each thread of control can be stored separately, and events from each thread can pass through the same controller.

Windowing systems typically have a separate thread of control for each application. Embedded systems have slightly different needs. Many process control devices consist of a settings area and a monitoring area. Allowing each one a thread of control helps reinforce in the user's mind that monitoring and control are separate activities. Consider the furnace interface in Figure 9.5. While the user is typing in a new temperature, he may decide to check the average temperature. He can press the Average button to see the display change. The number he entered on the keypad is still valid. He can return to that activity and accept the new value. If you had a single thread of control, operations on the monitoring side would cancel incomplete actions on the settings side, and vice versa.

As an aside, the Accept key in this control panel is deliberately kept as far as possible from the Clear key. These keys are often placed next to each other because they have related functions. Unfortunately, on many systems, the cost of pressing one instead of the other is high. In this case, the Accept key was placed next to the newly accepted figure to encourage the user to look at the digits just entered before committing them.

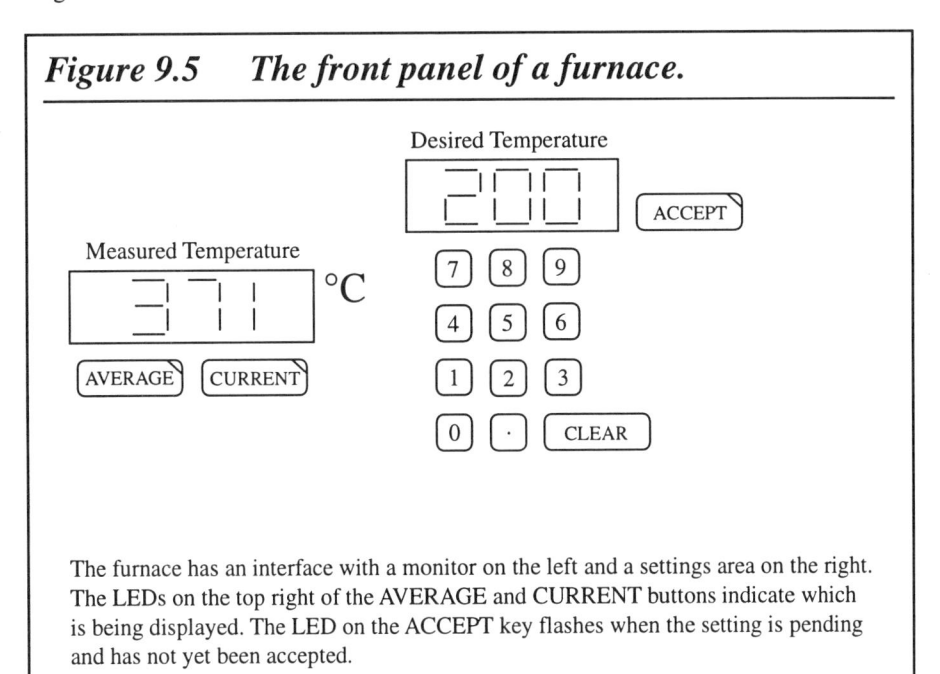

Figure 9.5   The front panel of a furnace.

The furnace has an interface with a monitor on the left and a settings area on the right. The LEDs on the top right of the AVERAGE and CURRENT buttons indicate which is being displayed. The LED on the ACCEPT key flashes when the setting is pending and has not yet been accepted.

9.2.8.3 Modes

Navigating some interfaces involves navigating through several modes. Sometimes, these modes correspond to the user's role or to a different phase of operation. In such cases, modes seem natural and necessary to the user. I expect my calculator to behave differently in hexadecimal mode then in decimal mode.

Modes have a bad reputation in the user interface community. This reputation was caused by user interfaces that made terrible use of modes. Text editors with insert and control modes create havoc when the user types a sentence and then realizes that he just typed seven commands. By then, half the screen has been indented and changed to upper case. My microwave oven has a grill mode and a microwave mode. I have often inserted some food and turned it on for five minutes, only to realize that I had the cooker in grill mode rather than microwave mode. The problem is not with the modes themselves, which are often necessary. It is with the means of recognizing which mode the device is in. Sometimes, the clue is small and easy to overlook. My VCR has a play mode and a record mode. There is never any confusion. The word *Play* or *Record* appears on the large, one-line text display when one of these modes is active. I cannot overlook this visual clue because there is so little other information presented.

The most troublesome modes are often invented solely to allow some controls or displays to be multiplexed. This is true for the text editor, which would be more useful if more buttons dedicated to the command functions were available. If you lack space for extra controls or displays, you may not be able to justify much space for a mode indicator. Now you have two bad features; a mode that does not reflect a task or role to the user, and a mode that is not obvious.

Use modes where necessary, but always make the current mode obvious.

9.2.8.4 Equal Opportunity

You should always keep the paths the user has to follow short and simple. One way to do this is to use data already available from the device output. When you select cruise control in a car, you do not have to dial in the speed. It is already available. The interface takes on output — the current speed displayed on the speedometer — and inputs it to the cruise control system. A VCR can use a similar mechanism. If the user sets a timer, the current channel can be assumed to be the channel used for this setting. The user may change it, but guessing defaults in this way reduces the number of interactions the user has to make.

The principle of using a piece of output as an input is known as *equal opportunity* (Thimbleby, 1990). It is common on desktop windows systems in which the text from one window can be cut and pasted into another. Many systems limit its usefulness by not allowing it for all text. I occasionally get an error message that I wish to e-mail to the system administration team. I am forced to retype the message because my computer will not allow me to copy and paste from a pop-up window into my e-mail window.

> **If data is already available, do not force the user to enter it again.**

9.2.8.5 Multiple Paths

If your interface provides more than one way to perform a function, ask yourself, "Is there a reason for each path?" You should ask this question, because the user will. If one way is slow but obvious, and another way is quick but only likely to be known by the expert user, then each has a purpose. The quick method is a shortcut. In other cases, you may wish to have features available from several modes, to reduce the user's need to change modes to find popular features.

However, if the alternatives are arbitrary, the user may assume that there is a difference in the result, a side effect that he never noticed while using the feature. This may lead to discomfort because the user believes he does not understand the interface.

9.2.9 Migrating from Mechanical Controls

Complete software control over the interface removes the user by one more level from the mechanics of the device. If software interprets all inputs and controls the device based on those inputs, then you have a *fly-by-wire* system. This affects the user in subtle ways. A dial connected to a potentiometer may have had a logarithmic relationship with the output, and the software control system may make that relationship linear or vice versa. Analog needles that show values are usually damped to prevent oscillations, that make it difficult to read the value. A real-time graph of the value shows variability not previously visible to the user. The user may believe that the new device is doing a worse job of controlling the process when, in fact, it is doing a better job of reporting the state of the process.

Having software in the loop greatly increases the amount and types of information that can be presented to the user. Engineers favor a more-information-is-always-better philosophy because of the nature of their work, but this does not always lead to better interfaces. The user wants enough information to solve the problem. He may not be as skilled as an engineer at filtering out numerical information that is only of peripheral interest.

When a display is converted from a mechanical indicator to a software-controlled display, you may be tempted to change the type, as well as the quantity, of information. Consider the fuel level indicator on a car dashboard. If you replace this with a small LCD display, there are several possible ways to present information to the user. You can express the remaining fuel as a percentage, in gallons, or in number of miles remaining before the fuel runs out. In each case, you allow the user to form a conceptual model of the fuel tank. Each model is different. But once you choose one, you must be sure you can fill in all the information the model requires. To convert from gallons to miles, you need a conversion factor. Using the car's average fuel consumption may not be sufficient if the driver has been sitting in a traffic jam for 45 minutes. What seemed, initially, to be a simple requirement now demands that the fuel monitor also monitor consumption. The user now has a new rule to learn. The number of miles that he reads from the monitor is valid only if he continues his current driving pattern. You may think this is trivial, but you should realize that every rule created by the designer must eventually be learned by the user if he is to make the most effective use of the interface. The rule may be learned from the user manual if he is one of the few people who read that section of the car manual. Otherwise, the rule is learned from careful observation, because the interface does not explicitly tell him. Such hidden rules can be dangerous. If there are too many of them, the user constantly finds himself surprised by the device's actions, and this leads to mistrust.

Having moved from an analog device to a digital display, many interfaces fall into the trap of giving the user more precision than he requires. There is little value in telling users that they have enough fuel for another 31.7 miles; since they would be foolish to wait until the last 0.7 mile before refueling. The vagueness of the needle in this case is an advantage. The needle says, "I don't know exactly how much further you can go. You are not at the panic stage yet, but if you see a gas station, you may as well stop." Because no one can know exactly how many more miles the fuel will last, this information is more appropriate than "31.7 miles." The more precise value may lure the driver into a false sense of security. A tempting alternative is to present the driver with a range of values. However, I do not think many drivers would appreciate a display that says 31.7 miles ± 10 percent. Remember, they are not all engineers and are not exposed to the concept of tolerance every day.

If the car is low on fuel, the driver should fill it up at the next opportunity. Needle gauges are notoriously nonlinear, but drivers rarely complain. The designer should copy the functionality of the needle. This is less threatening for the new user because he can relate easily to his past experience with the needle display. An interface controlled from software can control a needle, use a bar graph on an LCD display, or use a bar graph made up of a number of LEDs. Which appearance is chosen is less significant than the functionality attached to it. Users quickly adapt to a new appearance but take longer to adapt to changes in functionality.

9.3 The Graphical User Interface

Several years ago, adding graphics to an embedded system was a serious undertaking in both time and money. This is no longer the case. Processors are more powerful. Third-party toolkits allow developers to side step many of the more difficult parts of their programming. Flat LCD screens are cheaper and more robust. Touch screen interfaces are an input device that match the graphics output of desktop systems, with no need for a mouse and its awkward dangling tail.

Chapter 5 discusses several programming challenges raised by graphics development. This section looks at the usability issues involved when a user employs a graphics display to interact with the interface.

The ability to reconfigure the display at run time adds a level of interface flexibility that was not available before. With a custom display, one layout of buttons and displays had to be designed and evaluated. With a GUI, you have no limit to the number of possible layouts. Making each one user friendly while keeping it consistent with the others, is challenging. The payoff if you do it well is an interface that looks simple to the novice, but contains enough sophistication to satisfy the expert user. The interface can always display only the required information while hiding superfluous information unless requested.

9.3.1 Disadvantages of the GUI

Although a GUI has many advantages over custom displays, you should note the disadvantages. While a GUI allows several different controls on the screen, they all have the same tactile feel when you make an input. If the input is via a touch screen, they all feel flat. If the input is via a trackball, you use the same roll-and-click motions to manipulate all of them. With custom controls, a throttle controlling an aircraft's speed is physically larger and has a heavier feel than a radio's volume dial. The size of the control communicates the action's significance to the user. Imagine trying to drive your car with a mouse and screen as the only controls. You get the idea of how the feeling of control can be lost.

You can lay out custom controls in positions that fit the function. If you have an eject button on a VCR, it is intuitive to place it next to the slot through which the tape emerges. If a GUI is the only means of controlling the device, all controls must appear on that display, which means they are further from the related hardware. Another disadvantage of the GUI is that space does not usually permit the important controls to be permanently visible. This is not acceptable if the device is used in a situation in which the user may need emergency access to certain controls, or where some monitored information must always be visible.

A related problem is that if only a GUI is used, it is not possible to have all the controls visible at all times. Thus, the user may have to explore the interface to find some functions. The user may not choose to explore the interface unless he believes that there are functions that he has not yet discovered. With all the controls visible, the user will more likely say, "What is that for?" or, even better, "That dial must control the time delay," if the dial's purpose is obvious by its location and labeling.

Many embedded products get the best of both worlds by adding a graphics screen to support peripheral information at the same time that the most important user interactions use custom controls. This option allows little-used modes, such as configuration modes, to be implemented on the GUI alone, while normally running both GUI and custom controls. While the user manipulates the custom controls, information related to the changes is displayed on the graphics screen. For example, as the water flow in a pipe is adjusted on a dial, a diagram depicting the tank could show the water level rise and fall as the user turns the dial up and down. Such graphics are particularly useful for novice users who are constructing a conceptual model of the system.

9.3.2 Getting the Most Out of a GUI

Now that you understand how to make your interface graphical, you must still design intuitive interactions. Cooper (1995) discusses many design issues that concern the graphical designer in the desktop domain. Most of his advice also applies in embedded systems. Be warned that designing a screen layout is more suited to an industrial engineer or a graphic designer, although many programmers must perform this task. Most organizations do not see usability and interface design as concerns separate from software engineering. You should treat this as an opportunity to learn a new skill rather than bemoan the shortsightedness of management.

Once you decide to place a screen on your product, you must decide how it will be used. Some screens will be output only. The screen provides only information, possibly related to changes made on the custom controls. Such a display is not really a GUI because it provides only one-half of an interface, which is generally a two-way dialog. Several input devices are possible. The simplest is a line of buttons at one or more edges of the screen, with labels printed on the screen next to the buttons. A technique commonly employed on automatic teller machines. In that situation, the thickness of the glass that protects the screen can make it difficult to judge if each label matches each button. This offset is known as *parallax*. Figure 9.6 shows two views of an automatic teller machine, demonstrating this problem. This is an issue for a display that is behind the plane of the rest of the front panel.

To allow the selection of an area of the screen you need a pointing device. Mice are ubiquitous on the desktop, but they have disadvantages in embedded systems. The cable connecting the mouse to the device may get caught up in nearby moving parts. If the mouse is wireless, perhaps employing infrared technology, it is more prone to getting lost. The mouse needs a flat resting area where it can be manipulated. If such an area is not available, it can be mounted on the device or the top of the device itself can be used. However, the flat surface then accumulates manuals, charts, and abandoned cups of coffee. Laptops provide mouse alternatives that still allow a pointer to be used on the screen. Trackballs and pens are two promising alternatives, although the pen has the same disadvantages of the mouse in that it either has a tail or can easily get lost. Touch screens are a popular choice and are covered in a later section.

Some simpler forms of input include arrow keys, and less-conventional controls, such as voice or foot pedal. You can also track the movement of the user's eye to place the pointer wherever the user focuses his gaze.

Having decided on the mechanical form of input, you must decide the level of interaction. It can simply be a text interface, with text labels for the mechanical buttons. If a pointer is used, the sensitive areas of the screen can contain text. Text menus might be implemented on such a platform. The next level of control is the one at which a pointer manipulates graphical controls, such as sliders, pull-down menus, and 3-D buttons. There are usually a standard set of controls that can be reused. Once a user learns a control, he recognizes it and immediately knows how to manipulate it. For example, a tick-box, as seen in Figure 9.7, provides a way of turning some attribute on or off. You should keep the number of controls used in an application low to minimize the learning curve for new users. When choosing the controls, which may be available from a third party or written by you, be wary of control mechanisms that

Figure 9.6   Two views of an automatic teller machine.

The view intended by the designers.

Lodgement
Withdrawal
Balance

The view for a shorter customer. The labels and buttons do not line up.

Lodgement
Withdrawal
Balance

Figure 9.7   Tick-boxes.

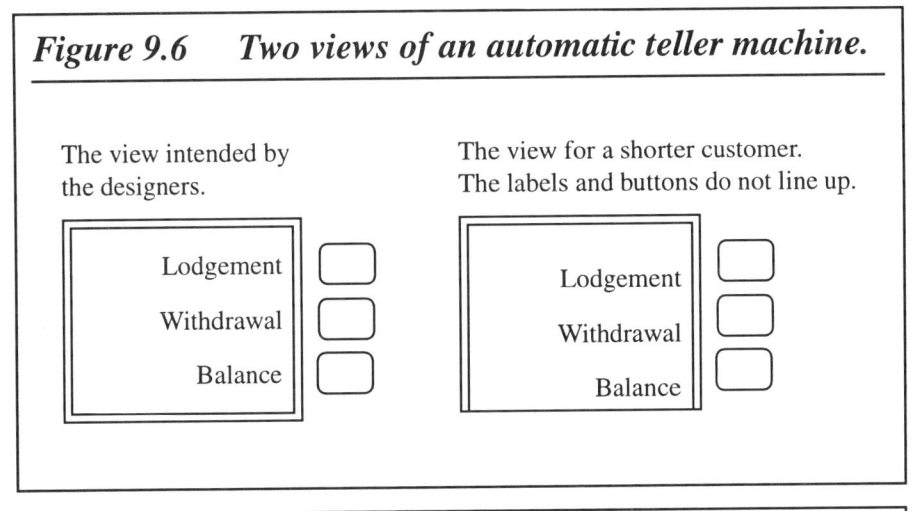

Key Clicks ☑

Screen Time Out ☐

Tick-boxes allow an attribute to be on or off. Key clicks are enabled while screen time outs are not.

work on the desktop but may not transfer to your embedded system. Double-clicking is hard to learn, especially if your target user is unfamiliar with a desktop computer. Dragging may be difficult with some input mechanisms, such as touch screens.

The next level of GUI development is to provide interactions in which the input and output is graphical by nature, not just a set of controls that could have been implemented with mechanical switches, dials, and sliders. Instead of outputting a numerical value, the value can be graphed over time, giving the user a better sense of changes within the process. A robotic arm might be controlled by moving a graphical representation of the arm on the display. This type of control is known as "a direct manipulation interface." When done well, it is more effective than the controls previously described, but it usually takes more programming effort to implement direct manipulation interfaces.

Direct manipulation is not necessarily a manipulation of the physical item's image. It is often a manipulation of a more abstract representation. Figure 9.8 shows a control for a flashing light, though this interaction might apply to any digital signal. If the user were required to enter the duty cycle as a percentage and the period in seconds, he would have to know what duty cycle and period mean. With a graphical control, the diagram makes it obvious what the change produces.

Although graphically representing entities to be manipulated is good, it is not always clear whether you want to imitate the controls that might be on the device had the GUI not been present. For example, if previous generations of a device had analog needle indicators to show temperature, should you graphically display a needle that sweeps an arc in imitation of the old control panel? This looks more familiar to the novice user than a bar graph. It is only a minor advantage, because users adapt quickly to changes in the appearance of information if the same information and functionality

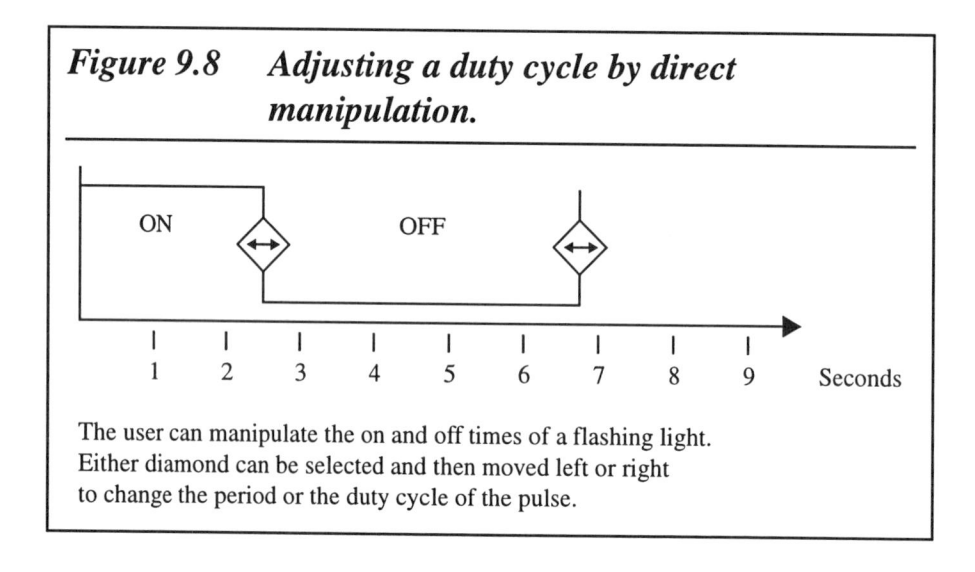

Figure 9.8   Adjusting a duty cycle by direct manipulation.

The user can manipulate the on and off times of a flashing light. Either diamond can be selected and then moved left or right to change the period or the duty cycle of the pulse.

are presented. You also lose real estate. A line graph can show several temperatures in the space occupied by a needle, as seen in Figure 9.9. More information may not be desired when you deal with novice users. In those cases, imitations of controls the user is accustomed to may be reasonable, but a more powerful interface is possible when the user is comfortable with the new technology.

9.3.2.1 Color

When I was in school, a standing joke, whenever a group of us gathered to ponder a new software project, was to ask, "Well, what color should it be?" Of course, they were software projects, so color had no meaning. Since then, graphics has become such a large part of the software world that most of us have had to decide what colors to use. Again, this is an area in which some professional graphic design advice is useful — the number of horrific home pages of software engineers proves that some of us should be banned for life from holding a paint brush. Do not use every color under the sun. Video games use lots of color, but they intend to be visually striking or even shocking; however, for control applications, you probably want more subdued colors that hint rather than scream.

Color is best used as a redundant clue. If you have another visual indication of the group you are defining, you will not exclude the one in twelve males that suffer from color blindness. It also allows for black-and-white documentation — which may be an issue in cost-sensitive applications.

9.3.3 Windows

On the desktop, the GUI is synonymous with windows. You can assume that any GUI uses a mouse or other pointer device and provides many overlapping, scrollable windows that can run independent applications. The embedded environment does not have

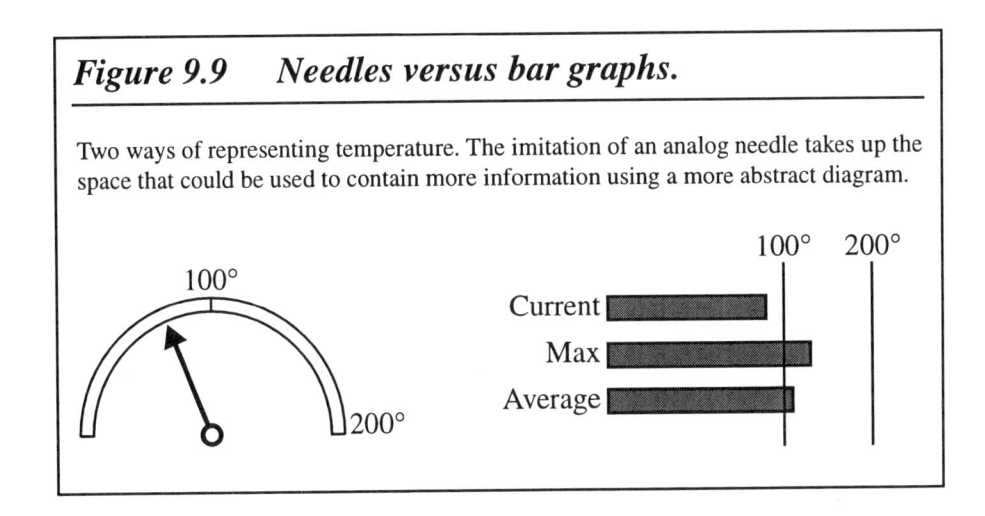

Figure 9.9 *Needles versus bar graphs.*

Two ways of representing temperature. The imitation of an analog needle takes up the space that could be used to contain more information using a more abstract diagram.

the same needs. One application may have several different states, each with a corresponding layout. I call these layouts dialogs. Generally, you have no need to allow the user to navigate arbitrarily from one dialog to another. The application can switch as the state of the system changes, or it can allow a button on certain dialogs to enable the user to switch to another dialog. The concept of a generic mechanism to minimize a dialog to an icon while the user uses another dialog does not apply.

With a window manager, the user experiences a learning curve as he establishes how to resize windows, move windows, and switch between active windows. Configuration work is overhead that does not contribute directly to solving the problem. Learning time is short for computer-literate users, but nontrivial for users who are unfamiliar with computers. Many embedded systems users consider themselves not computer literate. These people withdraw money from automatic teller machines, set the timer on a VCR, and cook food in a microwave. They do not consider that these acts involve computers. If they are introduced to equipment with a windowed interface, they view that equipment as a computer and may resist using it.

Another issue, if you want to use a window manager, is that frames for the individual windows take up real estate. If you use a small LCD screen, this real estate can be a big price to pay. The space allows the user to distinguish the window boundaries but does not provide useful information.

Some mechanisms can be borrowed from full-featured window managers. In a window manager, each window maintains its state when the user manipulates another window. Thus, you have one thread of control per window. An embedded system can allow the same power for different dialogs.

In cases in which you do want to use a full-scale window manager, ask yourself: "will important system state information be hidden because it is displayed on a window that is covered by another window?" On a desktop system, most changes on the screen occur in response to user input. However, on an embedded system, a lot of output may be in response to external events. You may need to use windows that pop-up in response to external events to notify the user. If this pop-up must be confirmed, it may interfere with normal work flow and can be annoying. An alternative is to devote a small portion of the screen to status and ensure that this area is always visible. The advantage of the pop-up approach is that the system knows whether the user has confirmed an event. You may need to alarm some events if the user does not acknowledge.

When discussing text menus, I have pointed out that people prefer shallow tree menus with many items at each level. Much the same issue arises with a GUI that has several dialogs. The more dialogs available, the greater the navigation problem. Once navigation becomes difficult, you must add more keys to provide shortcuts from one screen to another, using up valuable real estate. Try to make each dialog reflect a role or a phase of a task so that the user will be less likely to change dialogs.

It is easy to get sloppy in the layout if many dialogs have to be designed. One automatic teller machine that I use regularly has the screen shown in Figure 9.10. The customer has just asked for the balance. The balance is displayed in the middle of the screen, but the question, "Do you want any further transactions?" and the two answers are separated by the display of the balance. Reading this screen out loud tells you that something is wrong. After the user has read the question, the next thing his eyes should find are the possible answers, not unrelated information.

9.3.4 Pointerless

You can assume the desktop will have a mouse, but an embedded GUI may have to function with no pointing device. Keys may be aligned with one or more edges of the screen, allowing display of the key labels on the screen itself. These keys are sometimes called *soft keys* to indicate that their meanings can be changed by the software. If you take this approach, note that aligning the soft keys on the sides allows for longer labels versus aligning the buttons on the top or bottom edge. If the buttons must go on the bottom, staggering the labels, as shown in Figure 9.11, allows for longer labels, though it does take up a lot of real estate. This may work well when the soft keys are not often used. Note also that this layout does not suffer from the parallax problem as badly as soft keys that are aligned at the sides of the screen. There can still be a misalignment if the viewer looks at this display from the side. But the parallax problem is usually the result of varying user heights.

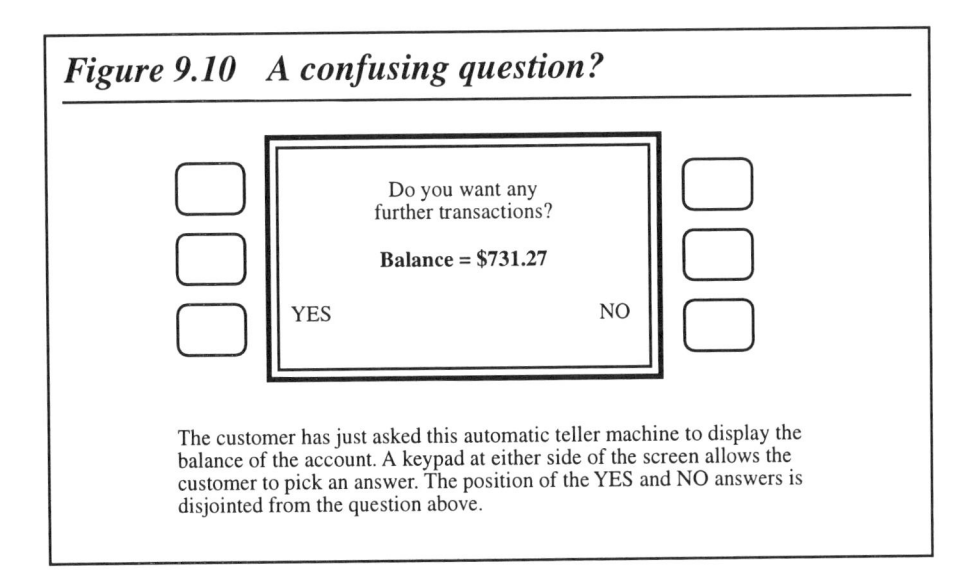

Figure 9.10   A confusing question?

Do you want any
further transactions?

Balance = $731.27

YES   NO

The customer has just asked this automatic teller machine to display the balance of the account. A keypad at either side of the screen allows the customer to pick an answer. The position of the YES and NO answers is disjointed from the question above.

You can also create keys with generic functions, such as arrow keys, which do not have to be aligned with the screen. In this case, you usually include a select, or accept, key that indicates that the currently selected item is the one you wish to apply. This is necessary because the arrow keys may traverse several options before reaching the desired one. As the user traverses the options, their images on the display change to show the currently selected item. The user does not want to activate the options as they are traversed, so a separate action, such as an accept key, is used when the desired control is reached. This is not an issue with soft keys as the user can directly select the desired option.

MS-Windows allows complete control from the keyboard. When you select items in a dialog box, you can use the tab key to jump from button to button or from item to item. You press the Return key when you reach the intended item. You can imitate this on embedded systems when no pointer is available. On Windows, this mechanism is rarely used because the visual indication of the targeted item — a transparent rectangle drawn with a broken line — is subtle. If this is the only mechanism you make available, you can afford to make the visual distinction of the target item more striking. In Figure 9.12, the background color of the target item changes. As long as you limit the number of options, the Next key can wrap around to the first item after the last item has been passed. You no longer need a second key to traverse the list in the opposite direction. An improvement on a Next key is a dial. As the user turns the dial, the target item changes. The Accept key applies the selected option. Once a dial is available, it can also be used for changing numerical values. You may want to place a LED next to such a dial to indicate times when turning the dial is valid.

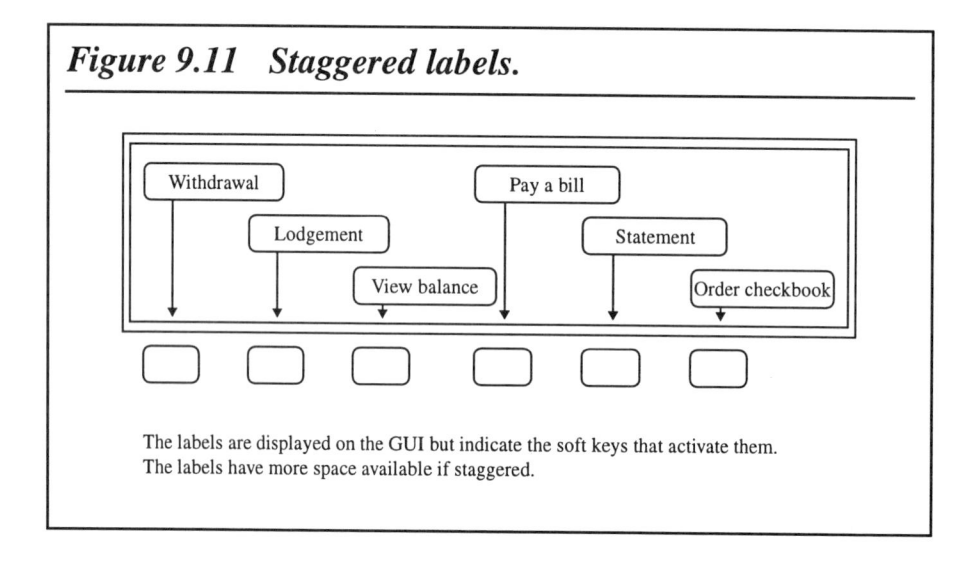

Figure 9.11 Staggered labels.

The labels are displayed on the GUI but indicate the soft keys that activate them.
The labels have more space available if staggered.

This scheme allows the selectable items to appear anywhere on the screen. This is useful if the items are placed on a diagram. If the screen displays a map, the Next key or the dial can step through each town on the map until the desired one is selected. This is not feasible with soft keys aligned at the side of the screen, because town names would have to be placed next to the keys rather than at their appropriate positions on the map.

9.3.5 Touch Screens

Touch screens are becoming a popular input device for embedded systems, especially in applications for which the walk-up user is likely to be a novice. Initially, the user must be shown that the screen itself is touch-sensitive. Users do not immediately guess this, because years of watching TV has conditioned them to assume that a screen is an output-only device. For a walk-up application, you may wish to display one button in the middle of the screen with the label, "Touch me to begin." This teaches the user that the screen is touch-sensitive, and also illustrates the visual hint used to indicate which parts of the screen are touch-sensitive — for example, a 3-D look on the buttons.

Sensitive areas of touch screens must be large if they are to accommodate stubby human fingers. A pen input allows more precision, but connecting a pen to the device is not always convenient. So the problem is that although graphics screens on embedded systems tend to be smaller than their cousins on the desktop, the space a control occupies is greater if you use a touch screen. This leads to a shortage of real estate on some interfaces and makes the investment in a larger screen worthwhile, even if the resolution does not increase.

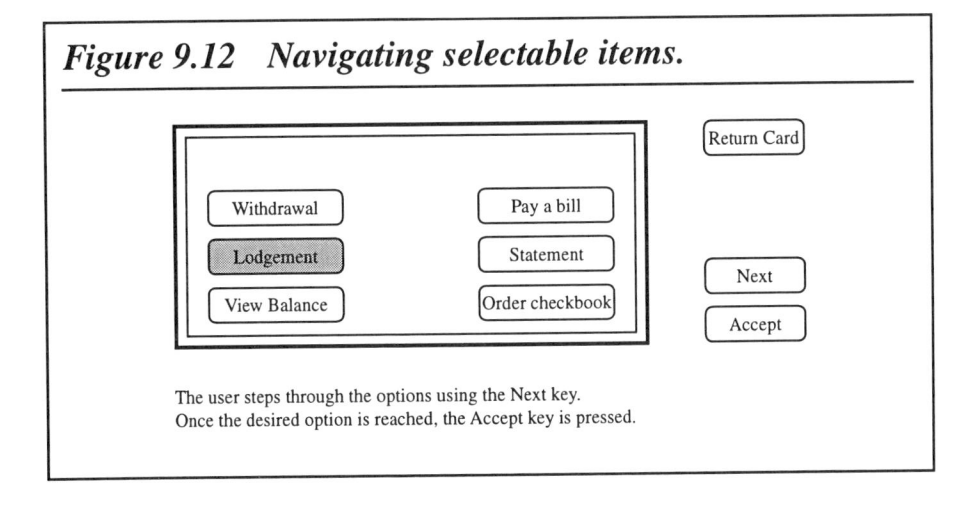

Figure 9.12   Navigating selectable items.

The user steps through the options using the Next key.
Once the desired option is reached, the Accept key is pressed.

Chapter 5 describes a scheme in which the sensitive area can be larger than the space occupied on the display. This may alleviate the problem of limited screen real estate. Once the display does become crowded, you should realize that vertical lists are harder to use than horizontal lists because the user's hand may slip down as it is removed from the display. This is most noticeable if there is no perch for the user to rest the heel of his hand. Even when you provide a heel rest, the knuckle bends as the user moves his finger and may possibly select an item lower down on the display. This is not an issue for horizontal lists, because the finger is unlikely to move sideways.

Some manipulations are more difficult on a touch screen than with a mouse because pixel-perfect finger motion is difficult. On one home automation project (Schneiderman, 1993), the initial version of the touch screen interface used clock faces to set timers. The user was required to touch a hand of the clock and rotate it to the appropriate time of day. Users found this motion difficult to control. Satisfaction with the interface rose when the clock faces were replaced with timelines on which times were marked with flags. The user could move the flags with a straight line motion, which is less challenging than a rotation.

You can combine touch screens with audible feedback and speech synthesis for visually impaired users. Such interfaces are particularly applicable to publicly-available information kiosks and automatic teller machines. A gesture such as running a finger along the bottom of the touch screen can indicate that the user wishes to operate in a *talking finger* mode. You can provide verbal feedback as the finger moves into a touch-sensitive area. Alternatively, each option currently available on the screen can also be made available via a *speedlist*. The speedlist is selected by placing a finger on the top left of the screen and dragging it down. As each item in the vertical list is touched, it is verbally announced. The speedlist does not require the user to navigate a two-dimensional area that he may not be able to see. Once the desired selection is reached, an off-screen confirmation button activates the selection. These and other techniques for making touch screen appliances accessible to the visually impaired are described in Vanderheiden (1996).

9.4 Usability and Safety

Safety is a system property that resides partly in the device and partly in the people using it. A passenger aircraft is safe only so long as the pilot does not decide to fly it into a building. The safety of the device depends on the device being sufficiently tested and on fail-safe mechanisms that protect the users and others when some component, either software or hardware, does fail. Given a safe system, the user can maintain that property only if he acts responsibly and can interpret the information displayed by the device. Because device integrity can be violated by people who disable sensors, ignore alarms, or use the device improperly, a system is only safe if a person takes ultimate responsibility. That is why you are unlikely to see unpiloted passenger aircraft in your lifetime.

Even though the technology already makes it possible, passengers simply will not put faith in a system unless someone with the appropriate knowledge and skill is prepared to put his life at the same risk experienced by the passengers.

The operator is often the scapegoat for a badly designed man–machine interface. An operator cannot be blamed for ignoring an alarm from a device that issues several false alarms per day or for misinterpreting information that is displayed in a nonintuitive fashion. You may assume that safety-critical devices are operated only by trained personnel. But how true is that assumption? Consider driving a car — a deadly weapon if not properly controlled. The law requires that you pass a test before you are permitted to drive. However, the law does not require that you drive only the car that was used during the test — the interface in another car may vary considerably. Hospital equipment is used by highly trained doctors and nurses. However, their training is mostly in the principles of medicine that apply to their patients. They are not necessarily trained to use every brand of equipment. You can expect the trained user to understand principles, but do not expect him to know that alarm volume is hidden under menu item 3 on your device.

Even a perfectly functioning device may contribute to a safety hazard if information is not presented intelligibly to the user. Three types of information are presented to the user: direct feedback to actions that the user has taken; monitored parameters of the system; and alarms that alert the user, usually audibly, to unusual patterns in the monitored data.

9.4.1 The Usability versus Safety Trade-off

Although it may seem intuitive that a device that is easier to use is safer, it is not always the case. To provide safety, you must often make the user work harder. In many situations, you have a direct trade-off between usability and safety. The Therac-25 was a cancer irradiation device whose faulty operation led to a number of deaths (Leveson, 1995). One of the safety features in the original design was that all the settings for the device had to be entered through a terminal, as well as on a control panel. This was seen as redundant by users of a prototype. It was redundant in the best sense of the word, but it was not appreciated by the users, who assumed that the safety of the equipment was beyond doubt. The design was changed before release so that the settings could be entered on the terminal alone. Once the settings were accepted, by hitting the Return key, the user was asked to confirm that the settings were those that were actually required. This confirmation was performed by pressing the Return key again. This extra step was considered a replacement for the original second interface.

Unfortunately users soon learned to press the Return key twice in succession; they knew they would be asked for confirmation. The two presses, similar to a double-click on a mouse, became a single action in the user's minds and no review of the settings was performed. Because of a bug in the software, some settings were not properly recorded. The bug was a race condition created because proper resource locking of the data was not exercised. Because the cross-check had been removed, the fault was not detected. This was a case in which the design was altered to favor usability, but the safety of the device was compromised.

It is fair to say that if the rest of the design had been sound, then removing the second set of inputs would not have been significant. But the point of putting a safety infrastructure in place is to allow for the times when something does go wrong.

Also, note that the later design was more susceptible to the simple user error of entering a wrong value. If the user has to enter the value once on two different displays, chances that the same wrong value will be entered are slim. The software would detect the mismatch and not apply either setting. Often, safety measures serve the dual purpose of protecting against device error and user error. In intensive-care medical ventilators, the pressure rise in a patient's lung is a function of the lung volume and the volume of gas added. A pressure valve opens at a fixed pressure limit. Once the valve is open, an alarm sounds and the patient is exposed to room air pressure in the fail-safe state. This protects the patient against an electronic or software fault that may deliver a large volume of gas. It also protects the patient from an accidental setting of 3.0 liters rather than the intended 0.3 liters.

9.4.2 Validating Input

Trusting that the input reflects what the user wants to do is part of the robustness property. If the user can easily perform irreversible actions, a mistake may have serious consequences. The system must make it obvious that an important action is about to take place. Asking the user to confirm an action is common, but not necessarily the best approach. Sometimes, making the action different from usual user actions is enough. A piece of medical equipment might have an "Apply to patient" button that would be pressed only when the adjusted settings are to be made active on the patient. This button can be placed a distance from the other input controls. Similarly, the power switch on a desktop computer is not placed on the keyboard.

In other cases, the system can attempt to guarantee that risk has been removed before proceeding. On paper-cutting machines used in printing shops, there is a risk that the blade that cuts the paper will drop while the user's hands are still moving the paper into position. You can remove this risk by placing two buttons on either side of the machine, which the user must press simultaneously to lower the blade. The user cannot press both buttons simultaneously if either hand is in the path of the blade.

Another side of this problem is that the device must know it is safe to continue operating in a certain mode in the absence of further instructions. Some devices employ a *dead man control*. A dead man control reverts to a safe state when force is removed. You can see these on devices as simple as a lawn mower. As the user pushes the lawn mower, he must hold a lever against the push-bar of the mower. If the handle is released, the mower stops. This reduces the chance that the user will place his hands or feet near the blade while the mower is running. In a case in which an incapacitated operator can lead to a dangerous situation, you should employ a dead man control.

While some situations offer a neat solution, safety often depends on a sequence of actions. The system should be capable of reminding users of incomplete actions. Pilots have been known to place paper coffee cups over a control they must revisit before they complete the current task. Try to make the interface's appearance different during a setup operation than during normal running so that an incomplete setup will be immediately recognizable. If the user leaves a task incomplete, you can use an audible tone to get the user's attention; he must either complete the sequence or cancel it to silence the tone.

Some equipment silently times-out by canceling an incomplete action. Time-outs guarantee that the device does not get left in a state in which one final key press acknowledges a change that was partly entered at some earlier time, possibly by a different user. However, this type of time-out presents several problems. The user may decide which step to take next when the initial steps are canceled. Estimates for the time it takes to read a screen may be optimistic if the designer assumes the user is a native English speaker. Even a time-out of several minutes is not long enough if the user decides to consult a manual or a colleague. The user may put the device into a state that is confusing. By the time the local expert sees it, the device has canceled out of the unusual state. Another danger is that the time-out may occur just as the user moves his finger toward the final key of a sequence. The user may not be sure which occurred first, the time-out or the completion of the sequence. He is then unsure whether the changes he input took effect.

Construct dialogs that are clear and unambiguous. On one device, I displayed the message "Press ACCEPT or CLEAR" when the user had been offered the choice between confirming or rejecting a change. However, the Clear key was to the left of the Accept key. This could lead to confusion as the user saw the word *clear* to the right of the word *accept* in the text message. See Figure 9.13 for an illustration. Note that this example also breaks the earlier rule that the Accept and Clear keys should not be near each other. That mistake was made, then the confusing text messages compounded the error. There is a danger that the first-time user may press the wrong key

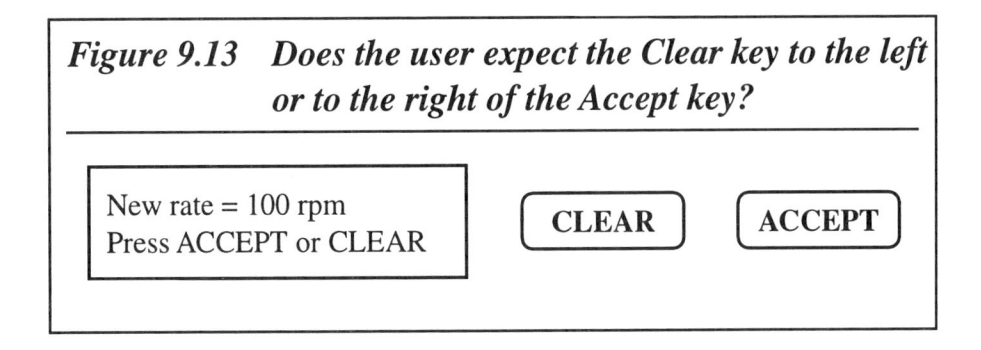

Figure 9.13   Does the user expect the Clear key to the left or to the right of the Accept key?

```
New rate = 100 rpm
Press ACCEPT or CLEAR        [ CLEAR ]    [ ACCEPT ]
```

once before learning the correct layout. Although the confusion caused here may seem slight, the greater danger is that when a user reacts to an emergency, he may follow the wrong impulse and press the wrong key. A state of emergency is obviously the worst time in which to make a mistake, but in this case, it is also the most likely.

9.4.3 Monitoring

Monitored values are the user's window into the process. The monitored values should tell the user the state of the system at a glance. It is not sufficient in a safety-critical system to have an alarm system that only detects whether the settings are being met by the system. One reason is that the set parameter may not be the appropriate value to monitor. A pilot uses the throttle to set the amount of power that the engine should consume. The air speed is monitored to inform the pilot of progress. Factors such as rate of ascent affect the air speed and cannot factored into the throttle setting.

Another problem when relying on settings and alarms alone is that alarms cannot react quickly enough. If you set alarm bands so tightly that a slight system anomaly causes an alarm to sound, you will see many false alarms. You then have the cry-wolf scenario in which alarms are ignored in emergency situations. If the bounds are set widely enough to avoid false alarms, alarms do not occur as early as desired when a parameter slowly drifts out of its desired range. The only solution is to provide nonintrusive monitored information that allows observation of the state of the system over time. Simple changes, such as a monitored temperature that rises slowly after the set temperature has been changed, reflect the proper operation of a furnace.

Values displayed in an analog fashion, such as a bar graph or an analog needle indicator, are quicker to read and better at providing relative measures, especially if minimums and maximums are shown. Digital displays are more precise but require more concentration by the user. Ideally, you should provide both. If average or recent values are available, the user can notice sudden changes in behavior. These may be early warning signs of an impending problem.

You must decide how much information to present to the user. You should avoid presenting too much information, which makes important patterns difficult to decipher. When safety is an issue, you should present information from as many independent channels as possible. Consider a furnace with two monitoring thermistors positioned in different parts of the furnace so that the user can monitor how evenly the furnace is heating. The user can look at the average or current temperature on each one. Figure 9.14 shows two possible displays for this furnace. Both have the same manufacturing cost because they use the same components. In both cases, each thermistor's average or current temperature is displayed. In terms of safety, however, display 1 has a distinct advantage. In display 2, if thermistor 1 fails and its data is unreliable, it is possible to display only information from the faulty sensor. However, if display 1 is chosen, data from each sensor is always visible. A faulty sensor will be noticed because the values from both sensors will diverge greatly. In any case, if the user monitors all four values regularly, the problem will be noticed. However, with display 2, the user may never bother to switch to the second thermistor. After a few weeks of using a properly functioning system, the user may believe that the variation between the thermistors is not significant for normal operation. A user's behavior always adapts to the normal situation, sometimes at a great cost should a fault occur.

You can argue that a simple alarm can check the difference between the two thermistors and annunciate if the difference gets too large. If the second thermistor is added for redundancy alone and is located in the same part of the furnace, then this may be feasible. In many control systems, however, separate sensors monitor points that are expected to be different for a number of reasons. It can be difficult to characterize the differences. During warm-up or cool-down, the differences may be larger than those seen during quiescent conditions. If an alarm's tolerance is set wide enough to cover all possible situations, it may be wide enough to allow a faulty sensor to pass its test. This leaves you back at square one, where the user must evaluate the information. Variations between the sensors may make the difference unsuitable as an alarm parameter, but the difference should be checked at some point when the system is stable, as a safety check that the components are functioning correctly.

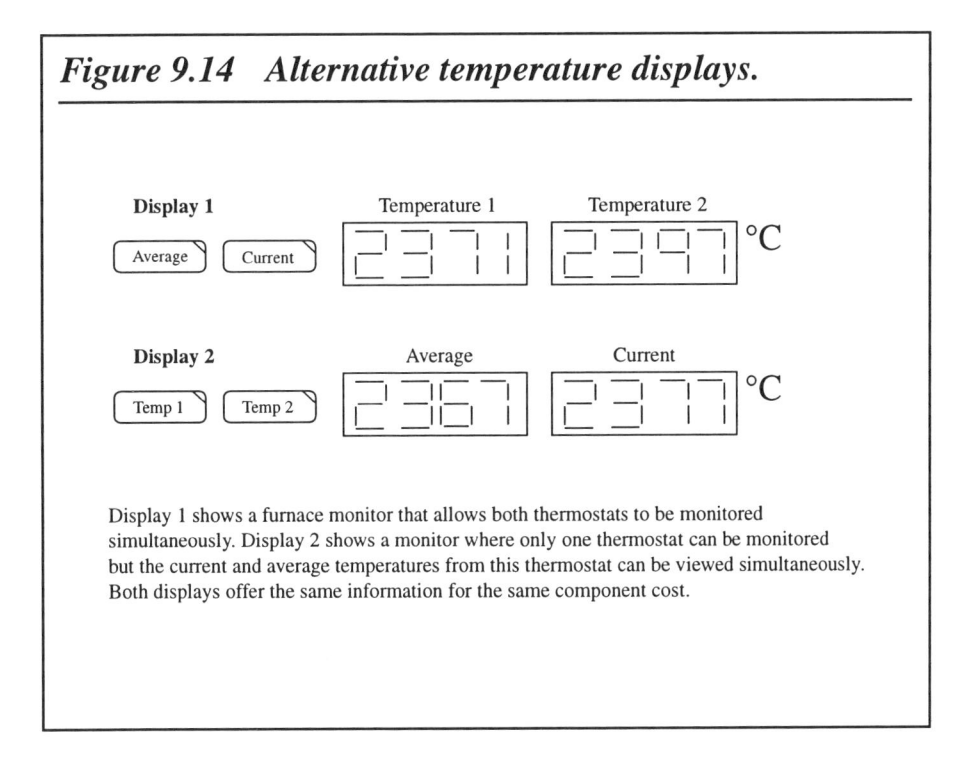

Figure 9.14 Alternative temperature displays.

Display 1 [Average] [Current]   Temperature 1 271   Temperature 2 297 °C

Display 2 [Temp 1] [Temp 2]   Average 267   Current 277 °C

Display 1 shows a furnace monitor that allows both thermostats to be monitored simultaneously. Display 2 shows a monitor where only one thermostat can be monitored but the current and average temperatures from this thermostat can be viewed simultaneously. Both displays offer the same information for the same component cost.

9.4.4 Alarms

Safety-critical systems require both an alarm system and a monitoring system to provide warnings of device failure and to warn of possibly dangerous patterns in the monitored parameters. If there is a possibility that the user is not looking directly at the device, you should include an audible alarm. For some classes of device, an audible warning is legally required.

In some cases, the monitor and alarm can be combined as a single display. Figure 9.15 shows a temperature bar graph labeled to show where the alarm is to be annunciated. Color coding can also be used. This type of display allows the user to anticipate the alarm before it occurs and to take evasive action.

9.4.4.1 Get Your Priorities Right

If there are many alarms, you should divide them into groups, based on their importance. Possibly only a subset of the alarms require audible annunciation. Different sounds and colored lights can distinguish the severity of the condition. Conditions can be distinguished by urgency and also by importance. An important condition may not need to be rectified immediately, but neither should it be put off indefinitely. A backup system failure falls into this category. Such alarms may get promoted in priority if the condition is not rectified.

Once several categories of alarms have been chosen, ensure that the alarms do not all migrate toward one category. As each developer considers his area to be of the utmost importance, each will be inclined to categorize the alarms in his area to the highest priority. Making all alarms fit into the highest priority category reduces the impact of a high-priority alarm because the user will see and hear it too often. The opposite can also happen. If the highest-priority alarm is invasive because it has a very loud sound, then few alarms should be placed in this category. This may indicate that the severity, in terms of audible annunciation, of the highest priority should be reduced.

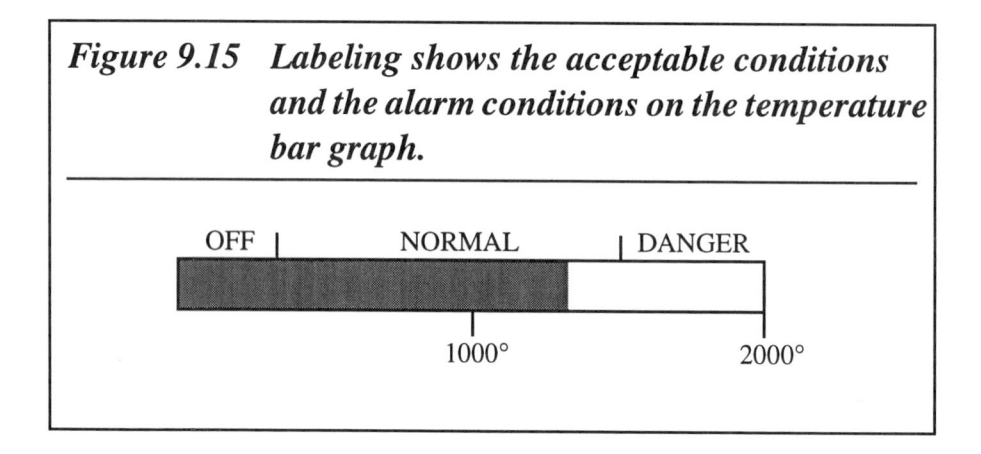

Figure 9.15 Labeling shows the acceptable conditions and the alarm conditions on the temperature bar graph.

When distinguishing the severity of alarms, the frequency of flashing lights is perceived as indicating degrees of urgency. For the sounds, urgency depends on speed, fundamental frequency, and number of repeating units. Sounds should not be so intimidating that the user's first reaction is to silence them rather than address the problem. However, factors include the user's work environment and how many separate pieces of alarm equipment are present.

Once alarms are categorized, you may need to assign them a priority order. This allows the operator to attend to the highest-priority condition first. If a list is formed, typically on a text or graphics display, you can order the alarms in chronological order. The order of alarm occurrences may give the user valuable information about the cause of the alarm. Ideally, the list is available in both forms.

9.4.4.2 Text Lists versus Dedicated Annunciators

Video display units (VDUs), whether LCD screens or more bulky cathode ray tube (CRT) screens, allow you to display an alarm as a text message on the VDU, rather than having a dedicated indicator. If the number of possible alarms is high — nuclear power station monitoring rooms have hundreds of possible alarms — then the VDU is the only option because of the amount of space that would be occupied by many individual annunciators. The VDU has an added advantage if secondary information, such as a value for the monitored parameter, can be displayed with the alarm. A VDU can also order the alarms, either chronological or priority-based. The user can be given several filtering options, such as displaying only temperature-related alarms or alarms that occurred in the last five minutes. Several views of the system can provide the user with extra insight into the cause of the alarms. Sometimes, the user tries to identify a real problem that is obscured by many nuisance alarms.

On the other hand, individual annunciators (often LEDs and a label), allow a visible pattern to emerge from the combination of conditions at any given moment. These annunciators are sometimes called *tiles*, since they may form a large area of small squares. There is no real estate competition between alarms; one alarm never forces another off the screen. The user knows which condition is associated with an indicator and does not have to read a lot of text to establish it. Text-based alarms on a VDU may cause confusion by giving the user too much information. It is too easy for developers to choose to give a long and informative description of the condition. In an emergency, many conditions are annunciated simultaneously, challenging the user's ability to read all of the generated text. An LCD-based VDU may not be bright enough for an alarm to be noticed from the far side of the room. Sometimes, audible alarms are not directional enough and the user may need to establish the alarm's source by glancing at many devices. This is particularly true in hospital intensive-care units, where each patient can be connected to several devices capable of generating audible alarms.

You can compromise by using bright indicators to indicate the level of the most urgent alarm, and then include further information on the VDU. Alternatively, you can use a small number of annunciators for the most important conditions while the less frequent or less urgent conditions can be displayed only on the VDU. Data suggests that individual annunciators outperform a VDU with text only. But VDUs surpass individual annunciators when they make use of color and graphics that allow the user to analyze the data (Stanton, 1994).

9.4.4.3 Cry Wolf

Many alarm conditions are not necessarily better than few. Nuisance alarms may be ignored or disabled. This is worse than not having the alarm at all. If the alarm is absent, you have to accept that the parameter simply is not monitored. If the alarm is present but unreliable, it gives the superficial impression that a dangerous condition will be detected. In practice, users may assume that an alarm is false, like many previous ones, and ignore it.

In some cases, the nuisance alarm did not indicate a failure but was simply specified to a tight criteria. Criteria that seem reasonable on the drawing board are not always so reasonable when manufacturing variability, real operating conditions, and years of wear and tear are factored in. An alarm sensor can be replaced or fixed if it fails, but if the design limits are wrong, the users may not be able to change them.

You can reduce the number of false alarms in several ways. First, pick reasonable limits for the parameter you are monitoring. This may not be enough, however. The signal may be noisy. If so, you can filter the signal at several levels. The analog signal can be electronically filtered to dampen some noise. The software can specify that the alarm must be active for a period of time before the alarm sounds. If the condition is checked on a regular basis, a number of bad readings can be required before declaring the alarm. A four-out-of-five readings rule may work better than a four-in-a-row rule, avoiding the situation in which a genuine alarm condition occasionally spikes into the valid area and prevents the alarm from sounding. Figure 9.16 illustrates this point.

Some alarms sound only in certain modes of operation. During startup or shutdown, conditions may occur that are not acceptable during normal running. Stanton (1994) calls these alarms *standing alarms*. You can avoid standing alarms by widening limits or disabling some alarms during the changeover period. Take care with this approach to ensure that the disabled alarms are not left disabled after the changeover is finished. It may not be wise to filter out all standing alarms. In some cases, users expect alarms in certain phases of operation. This reassures the user that the requested change is actually happening and that the alarm is functioning correctly. If there are few alarms and they do not disguise critical conditions, then allowing the user to observe the alarms may be better than filtering them out completely.

9.4.4.4 Testing Alarms

Alarms should be tested regularly during a device's operating life. Testing the alarm system during design can verify that the system is capable of detecting the appropriate alarm conditions. However, it is important that during the life of the product, alarm systems do not become ineffective because of component failure or unexpected environmental conditions. Other system features may be exercised so often that the user is soon aware if they fail. Alarms may get exercised rarely, but they must work when exercised. If alarms are not tested regularly and they fail silently, the operator may perform dangerous actions but believe them to be safe because no alarm occurred.

Two levels of testing are possible: external and internal. External testing is more reliable, but may not always be a viable option.

With the external method, the condition the alarm is meant to detect is physically created external to the device. For example, a match might be lit to generate smoke near a smoke detector. The external method tests the entire system so you can be sure the alarm will function properly when it runs normally.

With the internal method, you can only test certain parts of the alarm system as part of the system's self-test. It is possible that the alarm will appear to work in test mode, but could fail when running normally due to a fault at any point in the alarm system. An example is a button on a smoke detector that causes the alarm to sound. This ensures that the battery and speaker are working correctly. However, it would not detect a clogged air inlet. When you start the ignition of a car, the warning-indicator lights on the dashboard illuminate. This tells you that the indicator bulb is working. This test is worthwhile because the bulb is a likely point of failure. However, this test does not confirm that the entire alarm system works. If the temperature sensor used to detect an overheating engine fails it is not detected.

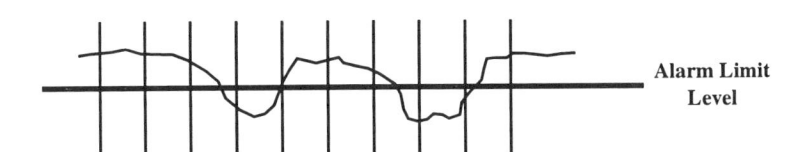

Figure 9.16 A borderline alarm condition subjected to filtering.

The vertical lines represent sample points. A signal above the alarm limit should cause the alarm to annunciate. This signal is above the limit most of the time, but a filter that requires four bad readings in a row would not trigger the alarm. A filter that requires four out of five readings to be above the limit would trigger the alarm, as desired.

Although the external method is obviously preferable to ensure reliability, in some cases, generating the condition is too hazardous or too expensive for routine testing. You cannot overheat a nuclear reactor to establish that the alarm systems are operating. The hazard would far outweigh the benefit of testing the alarm system. There are also cases in which the designers cannot depend on users to take the time to generate alarm conditions. In those cases, internal testing comes into play.

If you must use internal tests, keep track of components that are not checked by the alarm test. Make sure these components are covered by other tests and that redundant components and alarms ensure the system's safety even if the untested component fails. A furnace's temperature sensor may not be in the test loop of the overheating alarm because it is not feasible for the furnace to be overheated to generate the alarm condition. If the same sensor is used normally to monitor the furnace, temperature, and if it is regularly calibrated or cross-checked against another sensor, consider that component trustworthy; the fact that it is not involved in the alarm test may not create a risk.

If software controls the alarm level, the alarm level can be lowered to a point within the furnace's normal operating range. The alarm can then be tested without actually overheating the furnace. However, putting alarm limits under software control solely for this purpose is suspect and adds to the alarm system's complexity.

In some cases, you want to have several alarms that report the same condition in different ways. A motor can have one alarm with a limit set on the shaft's number of revolutions per minute. A separate alarm can check the electrical current drawn by the motor. If the motor is driven too fast, it is unlikely that this will go undetected by both alarms.

Avoid a false sense of security engendered by the assumption that you have redundancy when the alarm indicators are based on the same raw data. If the fuel-level indicator in your car stops moving, you assume it is broken. You would be unwise to wait for the low-fuel indicator light to illuminate before you stop to fill the tank. The failure may be that the fuel tank float is stuck in position. The needle indicator and the warning light depend on this float to function. A single failure can easily disable both indicators — so they are not redundant.

9.4.5 Discovering Safety-Critical Mistakes

Many accidents can be attributed to poor user interfaces. Accidents that involved devices in your industry may be well documented. However, such accidents may be rare enough that it is difficult to establish patterns, or the information may be heavily censored for liability reasons. In the most unfortunate cases in the aviation industry, the pilot does not survive to explain what went wrong. For these reasons, you should gather information on the near misses, especially if you can talk with the user and get a full understanding about what aspect of the interface caused confusion. This is not always a trivial exercise.

More experienced users are often happy to discuss situations that involved less senior and, possibly, less well trained staff. Your company's service and training functions may have many anecdotes of user mistakes. The biggest drawback with second-hand information is you cannot always find the person who made the mistake to investigate the actions he took.

Consider the situation in which you are part of a team that has just begun the development of a safety-critical system. Team members consider it important to discover where errors were common in the interfaces of your product's predecessors, especially if the errors were safety-critical. So you gather a cross-section of the user community and address them: "Please tell me about all the occasions in which you made serious errors using the current equipment. I am particularly interested if your mistakes put peoples' lives in danger." No matter how nicely you dress up this request, people will be reluctant to reveal accidents in which they were involved. In some cases, you will need to guarantee that the information will be confidential, especially from their employer. In the United States the Air Safety Reporting System (Leveson, 1995) provides a confidential means for pilots and others to report on aviation-related incidents in which safety was compromised. Confidentiality is a crucial contributor to the program, which has identified many risks in the aviation industry.

Chapter 10

The Future

The most satisfying thing about working on embedded systems is the feeling that you are building the future. I think most engineers are children at heart, who want to build part of the world they see on Star Trek. Embedded systems provide an incredible opportunity to affect everyday life in increasingly meaningful ways. I constantly meet programmers hired by companies that have manufactured electrical or mechanical equipment for decades but only recently hired their first embedded systems programmer because they just started to use microcontrollers to control their products.

This is good, as there is no shortage of work for us. It is good for industries that have mastered the art of using technology to scale their operations. On a modern assembly line, the effort involved in installing new technology is usually worthwhile and can be measured. The average consumer's view of all this technology may not be as optimistic.

Domestic electronics are ever-more baffling. We are bombarded by a combination of technological advances and marketing efforts to educate us with pseudo-scientific descriptions of products, in an attempt to convince us that our current collection of consumer electronics is obsolete and that we must buy again. If engineers cannot find time to read enough of the user manual to figure out the difference between Dolby and SurroundSound, then what hope is there for the less technically inclined?

If we cannot build VCRs and stereo music equipment that are simpler to use while still offering a full range of features, a new profession will emerge. A twenty-first century equivalent of the plumber is the person you will call to wire your TV sound so that it can be channeled through the stereo. He will reprogram the remote control so that changing the TV's volume will change the volume on the stereo amplifier proportionately. The car automatically opens the garage door on arrival, but it occasionally also opens the neighbor's garage door. Our handyman will soon put it right. He then gives the home computer, or home Internet appliance, a once-over. He might configure the security settings to ensure that the family's personal information is not accessible to the dozens of computers they communicate with each day. A quick check for new viruses shows that they are not in as much trouble as they were on his last visit. He then installs the latest versions of software that the family did not want to upgrade themselves, in case they broke something. Their Internet appliance is as central to their lives as the phone or mail, so they leave maintenance to a professional.

Would reliance on such a tradesman be a terrible thing? We accept that we need a plumber or an electrician occasionally, so why not someone to install and fix domestic electronics? The do-it-yourself enthusiast can still wade through the manuals and figure out how to do it for himself, but most people prefer not to worry about it.

Although this arrangement is feasible, it would represent a failure on the part of designers of devices that people should be able to pick up and use. Maybe a different route will be followed. The majority of people learn to drive, which requires an amount of specialized knowledge. Perhaps, as the devices around us get more sophisticated, so will we. This does not take away the burden of designing usable devices from the engineer. Many devices must still be usable by the occasional user — one who does not have time to climb the learning curve. Negroponte (1995) makes interesting reading for those interested in other opportunities our digital future has in store for us.

10.1 Working Together

Standards in the desktop world have made it possible to use many applications, right from startup. We can operate scrollbars, pull-down menus, and other controls because we have seen them in earlier applications. However, while standards have made interface components common, they sometimes limit the design. Standard components arrive in library form and are easier to use in a program. This encourages the programmer to use prebuilt components rather than define new components tuned to a particular application. This limits innovation and renders an application less usable. However, the usability of the family of applications that conform to this standard increases. From that point of view, standards make a big contribution.

The embedded world is making some inroads in that direction. The International Standards Organization (ISO) set standards for icons that represent the functions on a typical VCR. They have dictated the order of letters on a numeric keypad, although many current mobile phones do not conform.

These standards are bound to the way devices look and do not give much guidance about how devices should operate. For nongraphical front panels, it is difficult to legislate an interaction. Does a change require a completing action, such as an Accept key, or does it have immediate effect? Should arrow keys auto-repeat if held down, and if so, after how long? If there is a stop/go button with a single LED, does an on light mean that the device is running or that it is stopped? The light on my kettle tells me that it is on. My oven light tells me that it is on but that it has not reached the selected temperature. There are no easy answers when devices and their users are so diverse. Timers on the household appliances around me are controlled in different ways. Interfaces to many devices are just too diverse to be encompassed by a single set of rules that allows the knowledge I gain using a VCR to benefit me when operating a multispeed drill. On the desktop GUI, complying to a standard represents a large savings because libraries make it easier and cheaper to conform to the standard than to diverge. When front panel components must be purchased individually, migrating toward a common set is costly for the low-end devices. For this reason, standards across application domains simply do not happen on nongraphical embedded devices.

As devices become more capable of communicating with related devices, standards that dictate the way these devices work will have to evolve. If you are in a car on the way to the shop and you want to check whether there is milk in the refrigerator at home, your refrigerator must be able to answers the query, "Have you any milk?" in the same way as any other refrigerator. The refrigerator can detect the milk if it reads bar codes. But to receive and transmit the information remotely, it must speak the same protocol as your office computer or your car or even in such a way that the grocery store can detect that you are short and deliver it without asking. Right now, my TV remote does not talk to my VCR. This Tower of Babel will eventually convert into the embedded equivalent of the Internet, and when it does, the challenges to the user interfaces on those devices will increase. Do you add a control for each of the possible features on the devices with which you communicate? For such flexibility, graphics are the only option. But if your refrigerator's graphics have a different look and feel from those generated by your VCR, each new device will be a challenge to learn.

10.2 Graphics Everywhere

Graphics provide the only way to a vast range of features on a device that is limited in size. Many domestic devices will be controllable via the TV screen or another central domestic computer-controlled screen. The TV, incidentally, is a larger screen than most computer monitors, making it ideal for a touchscreen interface once the interaction goes beyond the capability of the remote control. Using one shared screen rather than one per device keeps the cost of having a GUI for a central heating system within reason. We are starting to see products that Web-enable embedded systems so that the graphical interface to the device can be any Web browser. This means that a graphical front end will have to be designed and programmed for many devices that, to date, have required only a custom front panel. It would be a dull designer, indeed, who elects to make the graphical front end a simple imitation of the front panel; although, as an option, an imitation of the front panel would be desirable for those already familiar with it. Graphical front ends allow for instruction modes that guide us through operations rather than making us hunt for a long-discarded paper manual.

Many graphics libraries have standards for powerful interfaces made up of components and controls that allow sophisticated desktop applications. I believe that another set of priorities must be applied to embedded devices in which ease of use outweighs the need for flexibility. Despite the disadvantages of graphics outlined in Chapter 9, economics and interoperability dictate that graphics will be with us for a long time. The embedded programmer's skill set must meet this challenge.

10.3 Voice Synthesis and Verbal Commands

The popularity of mobile phones, personal stereos, and palmtop computers demonstrates that we are willing to carry our technology around with us. That willingness will increase dramatically as the technology shrinks one more step and we can wear it on our wrists. We already have wristwatches with touch-sensitive displays. Once the device is always on us, many uses may emerge that, on their own, would not have justified a wrist computer/communicator. Say, you are in a shop and you scan the bar code of an item you are interested in. The watch can tell you that the price down the street is $10 lower. The information is accessed over a wireless network. In another example, two people want to exchange phone numbers and addresses — quickly done if the watches can communicate with each other via an infrared or other wireless link.

But what will shrinking computers mean for the programmers who create the software that drives their interfaces? Devices below a certain size make only very limited use of graphics for output. They are even more severely limited in the forms of input they can accept. Speech will become an important form of input and output for these devices. Some may think that no one will want to talk to his palmtop computer or to his digital watch in a public place. This will not be an issue — look at the number of people talking to mobile phones in all sorts of places. They could just as easily be accessing their electronic diary.

Voice input and output will also have a huge impact on automobiles. One obvious target is the old problem of trying to change channels on the radio while driving. In order to allow the driver to keep his eyes on the road, a steering-wheel-mounted button could activate voice-control mode, followed by a verbal command, "Radio on, channel 4, volume low." Other noncritical adjustments, such as opening windows and adjusting the air-conditioning, could be controlled this way. Voice for cruise control or brakes would probably be unwise. On the output side, synthesized voice has proved to be the most effective way of providing driver directions. It is more effective than a map displayed on a graphics screen. Again, the driver needs to keep his eyes on the road. Even if the driver does not need to look at the road, the information needed to take the next left is simpler to interpret than a map presenting every contour for 10 miles around. Synthesized voice has also proved successful in airplane cockpits, where wind shear detection and collision avoidance systems use speech to warn the pilot of rare, but urgent, events.

The use of voice and speech synthesis introduces several interesting challenges for the interface designer and for the programmer. The interface designer has to design an interface that may be controlled by someone who is within earshot of the device but possibly not looking at it. Difficult decisions must be made in deciding the tolerance levels. If a command has a 10 percent chance of being "Volume up," a 20 percent chance of being the command "Volume down," and a 25 percent chance of being a meaningless command, should the system turn the volume down and take a chance that it might be wrong occasionally? A type of acoustic debouncing is being applied. In some cases, that might be appropriate, but not on safety-critical systems.

There are fundamental problems, such as how to get the computational power required for voice recognition into a small embedded system at a reasonable cost. For the programmer, input timing is not as discrete as a key press. The voice command does not happen at an instant, but may take seconds to utter. If the context changes during that time, should the command be interpreted in the context that was current when the user started his sentence, or at the end? This could be an issue if the user uses voice controls, along with other forms of input.

10.4 Nuts and Bolts

After solving communication problems between devices and between the user and the device, the next big hurdle is mechanical. The hardest problems to solve in home automation are mechanical. There is no use being able to connect to and turn on the oven at home if the pizza is still in the freezer. How do you get the pizza from the freezer to the oven? How do you take the clothes out of the washing machine and put them into the dryer? Some of this can be solved by creating multi-purpose devices — changing the device instead of moving the item you want processed. Combined washer/dryers are already commonplace. Is a combined freezer/oven that far away? In some cases, we will still need to move things. The problem with performing such actions remotely is that you cannot see what you are doing. Someone may have eaten the last pizza an hour ago, or someone might have their hand in the freezer when the conveyer belt starts moving to transport our Italian delicacy. The user interface must provide enough feedback so that the user knows which actions they have instigated, and that they are safe. Currently, challenges like these are met in the control of robots on the surface of Mars and in probes into human blood vessels. As technology grows more affordable and the human–computer interactions suitable for remote manipulation become better understood, this technology will move into the mainstream. For more information on the potential of telerobotics, see Lavery (1996).

10.5 Software Quality

Currently, embedded software in many products enjoys the reputation of being robust when compared to the desktop world. When was the last time you heard a mechanic grumble that the engine control software crashes every time you shift down a gear and the lights are on? This will change as embedded software becomes more complex and closer to the software that runs on PCs. The set-top boxes that control our TVs may run applications from different networks — one program allowing you to buy goods seen on screen, another application for audience participation chat shows, another application to restrict or censor viewing programs or channels, another application to beep loudly and inform you that your favorite show has just started on the other channel. The list is endless. The combination of applications you choose may be unique and, therefore, untested. Fairly soon, you get into similar resource management issues that you have with PCs and workstations. Other devices will implement generic protocols to allow them to be controlled from a remote site or via a hand-held remote control that conforms to the shared protocol. The combinations of devices that will be able to work together cannot all be tested. This is where embedded software's reputation may start to crumble.

The lesson from the desktop world, if it could be called a lesson, seems to be that it is okay to write crummy software as long as you make it inflexible enough that your customers are locked in. Customers have to come back year after year, paying for upgrades that fix the latest set of bugs. There are times when software managers must weigh the possible damage from another missed release date and take the time to get it right. Getting it right seems to be only occasionally rewarded in the desktop application software industry.

This pragmatic approach may not work in the embedded world for several reasons. First, users tend to be less computer literate and, therefore, less tolerant of bugs, if they even recognize a bug in the product. Second, embedded systems often deal with fundamentally irreversible processes. If the microwave burns the dinner, no set of backup disks will bring you back to where you started.

Economics also dictate that the cost of software failure is much higher. Most embedded software vendors are really selling hardware controlled by software. If the software has to be upgraded, it is generally not possible for the customer to perform the upgrade. The manufacturer must bear the cost of labor for it. In the PC world, that labor is cleverly transferred to the owner. Because the vendor sells only software, the responsibility of performing the installation or upgrade is in the user's hands. Embedded software vendors are rarely afforded that luxury, and rightly so.

Interoperability between embedded systems inevitably leads to abuse, as we have seen on other large networks, such as the telephone system and the Internet. A friend of mine had a digital watch that could be used as a remote control for most makes of television. He would change channels in a bar or a restaurant. If remote control TVs can be abused this easily, the embedded world had better gear itself up for major security headaches when our alarm clock tells our car to start the engine and warm it up because it is a cold morning and his lordship does not want to have to face a freezing car in 10 minutes time, and by the way, unlock the doors so that he can get in quicker. Many other risks of our increasingly computerized society are explored by Neumann (1995). Quality software in the coming decades will not just have to be robust enough to keep the system up and running, it will have to protect itself against both accidental and malicious attacks.

10.6 Programming Activities

This book has been directed primarily at C programmers. Attempts to dislodge C as the language of choice in embedded systems have, by and large, failed. Even the U.S. Department of Defense is starting to neglect its strongest challenger, ADA, because it wants to leverage commercial off-the-shelf software. At the time of writing, the Pathfinder spacecraft is sending pictures back from Mars. Most of the software guiding that craft was written in C. For many embedded systems, C has problems, but it is difficult to find a better alternative. Java is currently touted as the next big thing in embedded system programming languages. I am not holding my breath. The virtual Java machine means a big overhead and an environment that isolates you from the hardware rather than providing controlled access to it. Java may find limited success on those few embedded systems in which it makes sense to run third-party software. Otherwise, it will not live up to the hype.

As embedded systems software becomes more complex, more of the software engineer's time is spent on the higher levels of software. It will no longer be sufficient to control all the hardware directly. The tasks performed by the typical embedded system, particularly in communications, will become more complex and involve several layers of software, which keeps much of the code far removed from the hardware.

As more embedded systems use off-the-shelf components, embedded PCs, and other single board computers, the software engineer will no longer spend long hours testing digital electronics. Tools will also improve and, in some cases, third-party libraries will be available to control some peripheral hardware. This will leave the engineer more time to work on the application domain of the problem the device is trying to solve. Hopefully, some of that work will translate into front panels that are easier to use.

Appendix A

RTOSs and μC/OS

This appendix gives enough information about real-time operating systems (RTOSs) and μC/OS that you can follow the example code and discussions in this book. For a full description of μC/OS, see Labrosse (1992).

A.1 Real-Time Operating System Concepts

Embedded devices often must do several things in parallel, or at least appear to do them in parallel. An RTOS makes this possible by sharing CPU time among several tasks. Each task is similar to a program; each has a separate stack and a separate thread of execution. Unlike the processes implemented on larger operating systems, RTOS tasks usually occupy the same address space. Thus, global variables are visible to all tasks and code may be shared between tasks.

Each task usually consists of an infinite loop that iterates through the task's responsibilities. This is sometimes called the *main loop* of the task. Many RTOSs can create and destroy tasks dynamically while the system is running. But more typically, the number of tasks is static. A timer interrupt drives the RTOS. At a certain timed interval, the RTOS is interrupted and must decide whether the currently running task can continue running or if a different task should be made active. If a different task is made active, this is known as a *context switch*. A context switch may also occur when an operating system call changes the status of a task.

The tasks may communicate by exchanging data in global variables, but access to global data cannot be atomic. One task could be accessing global data when an interrupt causes a context switch. Another task may now read the data in an inconsistent state. Similarly, access to hardware may need to be exclusive. Data or hardware that needs such protection is called a *resource*. A resource may be locked by a task for the period of time it is being accessed, then released when the resource is no longer being accessed.

A.1.1 Queues

Queues are another way in which tasks communicate. Queues are an asynchronous form of communication. One task sends information to another task, which the receiving task may not examine for some time. Neither task needs to wait on the other unless the queue is completely full or completely empty. The queue can be any length. The queue's width is the size of the messages contained on the queue. The RTOS provides functions to write messages onto the queue and to read messages from the queue. The RTOS guarantees that the queue is protected from simultaneous access. If a task attempts to read an empty queue, the task may choose to block and wait for a message to arrive. A blocking task is removed from the list of tasks that are ready to run until a message is placed on the queue. This blocking mechanism avoids the need for a task to continuously poll a queue, waiting for a message to appear.

A.1.2 Priority

There are many algorithms for deciding which task gets a portion of the CPU time. Some algorithms depend on the active task voluntarily giving up control of the CPU. Others share time equally in cycles. For the purposes of this book, I assume priority-based scheduling.

In priority-based scheduling, a numerical priority is assigned to each task. The highest priority task runs as long as it can. Lower priority tasks get to run only when all higher priority tasks are blocked and waiting on an event. This may suggest that a low-priority task never gets a chance to run, but in practice, most tasks spend a lot of time waiting on external events, on a queue, or on a timer. A well-designed system allows all tasks to run enough to get its work done. Idle time — when no task is doing any useful work — is a good measure of how much processor bandwidth is still available. If the idle time is zero, one or more tasks may never get a chance to run.

Priority should not be chosen based on the importance of the task. One factor to consider is how soon the task must respond to events. A feedback loop for a highly responsive circuit should have a higher priority than a task that responds to a human operator's relatively slow key presses. Another factor is how much work the task does before blocking again. If a task is busy for several seconds, it may prevent lower priority tasks from performing work that may require only a small slice of time. So tasks that run for a long time without blocking should be given a low priority.

A.1.3 Reentrancy

If you write code that uses interrupts or an RTOS, you must be aware of reentrancy. A piece of code is reentrant if it can be called from one thread of control while it is running from another thread of control. The simplest way to keep code reentrant is to ensure that all data it manipulates is on the stack. If it is called again, the second invocation will have its own stack frame to manipulate and there will be no conflict. If static data or hardware must be accessed, locking access ensures that the code remains reentrant. If you receive a third-party library and you intend to use it with an RTOS, you must find out whether these library functions are reentrant, because you may access the library from many tasks. Often, libraries written for PCs are not reentrant. If part of the code is not reentrant, you may have to make the restriction that only one task is allowed to use that piece of code, or you may have to add resource locks around all accesses to the code.

Printing functions used in the sample code are nonreentrant. This does not cause a problem because only one task is allowed to call them. Similarly, routines to read the keyboard are not reentrant, but only the keyboard polling task uses them.

A.2 µC/OS

µC/OS is an RTOS that implements these and many more features. The companion disk contains a compiled version with limited capability, built for the small memory model on a PC. The complete RTOS is available in source code format in µC/OS: The Real-Time Kernel (Labrosse, 1992).

A.2.1 Start-Up

The examples in this book initialize the RTOS with code of the following form.

```
/* Prototypes of task entry points */
void far        mvcStart(void *data);
void far        keyPollStart(void *data);
UBYTE           G_mvcStk[TASK_STK_SIZE];
UBYTE           G_keyPollStk[TASK_STK_SIZE];
UBYTE           G_mvcData;
UBYTE           G_keyPollData;
OS_EVENT        *G_mvcQueue;
void main(void)
{
    G_oldBP     = _BP;              /* Save current SP and BP      */
    G_oldSP     = _SP;
    G_oldTickISR = getvect(0x08);   /* Get MS-DOS's tick vector    */
    setvect(0x81, G_oldTickISR);    /* Store MS-DOS's tick to chain */
    setvect(uCOS, (void interrupt (*)(void))OSCtxSw);
    OSInit();
    G_mvcQueue = OSQCreate(G_mvcQueueData, Q_SIZE);
```

```
    /*
    Create the 2 tasks. Last Argument is priority.
    Note that the lower the number the higher the priority
    */
    OSTaskCreate(keyPollStart, (void *)&G_keyPollData,
                 (void *)&G_keyPollStk[TASK_STK_SIZE], 20);
    OSTaskCreate(mvcStart, (void *)&mvcData,
                 (void *)&G_mvcStk[TASK_STK_SIZE], 30);
    OSStart();      /* Start multitasking      */
}
```

The first five lines of main() are specific to DOS. The stack and base pointers are stored so that they can be restored to these values when you want to stop multitasking and return to DOS. The BIOS' old tick ISR is stored in G_oldTickISR, where it can be accessed when you want to restore the ISR to exit multitasking. The BIOS' tick ISR is then relocated to INT 81H, where it is called by μC/OS' tick ISR. The context switch vector is then installed at INT 80H. Use this interrupt routine to manipulate the stack pointer and the program counter to change the thread of execution from one task to another. Note that the μC/OS tick ISR is not yet installed, because you do not want to call this ISR until some tasks are ready to run.

OSInit() is called to initialize all data structures before any queues or tasks are created. OSQueueCreate() creates an empty queue of size Q_SIZE. OSTaskCreate() creates the task and is passed a pointer to the function that executes the task's main loop and pointers to the stack space that are used by each task. Note that you can allocate a different amount of stack space to each task. The address of the end of the array is passed in as the start of the stack, because the stack grows down in the PC architecture.

When OSStart() is called, the created tasks will start running, and this function will never return.

One task must install the tick interrupt. The following code runs at the start of the keyPollStart() function:

```
OS_ENTER_CRITICAL();
/* Install uC/OS's clock tick ISR*/
setvect(0x08, (void interrupt (*)(void))OSTickISR);
OS_EXIT_CRITICAL();
OSTimeDly(1); /* Synchronize to clock tick */
```

The OS_ENTER_CRITICAL() and OS_EXIT_CRITICAL() macros enable and disable the processor's interrupts, respectively. The call to setvect() installs the μC/OS tick ISR.

When a program using μC/OS exits, you must restore the stack and base pointers to the values they had before the program began multitasking. This allows for a clean exit to DOS. If these lines are not executed, the DOS shell may be corrupted and may crash when it starts to run another program. The following lines perform the tidy up and should be executed just before the call to exit() in any program that uses µC/OS.

```
OS_ENTER_CRITICAL();
setvect(0x08, G_oldTickISR);
OS_EXIT_CRITICAL();
clrscr();
/* Restore old SP and BP */
_BP = G_oldBP;
_SP = G_oldSP;
```

The following brief reference section gives a summary of the µC/OS functions used in this book's examples and provides a function prototype.

A.2.2 OSInit()

OSInit() is used to initialize µC/OS. OSInit() must be called prior to calling any other functions provided by µC/OS.

```
void OSInit(void);
```

A.2.3 OSQCreate()

OSQCreate() creates a message queue. A message queue allows tasks or ISRs to send pointer-size messages to a receiving task. The meaning of the messages sent are application-specific.

```
OS_EVENT *OSQCreate(void **start, UBYTE size);
```

start is the base address of the message storage area. A message storage area is declared as an array of pointers to void.

size is the size (in number of entries) of the message storage area.

OSQCreate() returns a pointer that can be used as a reference to the queue in calls to OSQPost() and OSQPend(). If no more queues can be allocated, the function returns NULL.

A.2.4 OSQPend()

OSQPend() is used when a task desires to receive messages from a queue. The messages are sent to the task either by an ISR or by another task. The messages are pointer-size and their use is application-specific.

```
void *OSQPend(OS_EVENT *pevent, UWORD timeout, UBYTE *err);
```

pevent is a pointer to the queue from which the messages are received. This pointer was returned to your application when your queue was created [see OSQCreate()].

timeout allows the task to resume execution if no message is received from the queue within the specified number of clock ticks. A timeout value of zero indicates that the task wants to wait forever for a message. The maximum timeout is 65,535 clock ticks.

err is a pointer to a variable that holds an error code. OSQPend() sets *err to either OS_NO_ERR if a message is received or to OS_TIMEOUT if a timeout occurs.

If a timeout occurs, the function returns NULL. Otherwise, the message is returned as a pointer to void.

A.2.5 OSQPost()

OSQPost() sends a message to a task through a queue. A message can be sent by an ISR or by another task.

```
UBYTE OSQPost(OS_EVENT *pevent, void *msg);
```

pevent is a pointer to the queue that holds the message. This pointer is returned to your application when your queue is created [see OSQCreate()].

msg is the message sent to the task. msg is a pointer-size value and is application-specific.

OSQPost() returns OS_Q_FULL if the queue is already full; otherwise, the value OS_NO_ERR is returned.

A.2.6 OSStart()

OSStart() starts multitasking. OSStart() must be called after you call OSInit() and create at least one task.

```
void OSStart(void);
```

A.2.7 OSTaskCreate()

OSTaskCreate() allows your application code to create a task that is managed by μC/OS. Tasks can be created prior to multitasking [before calling OSStart()] or by a running task [after calling OSStart()]. A task cannot be created by an ISR.

```
UBYTE OSTaskCreate(void (OS_FAR *task)(void *pd), void *pdata,
                   void *pstk, UBYTE prio);
```

task   is a pointer to the task's code.

pdata   is a pointer to an optional data area that can be used to pass parameters to the task when the task is created.

pstk   is a pointer to the task's top of stack. To determine the stack size, you must know how many bytes are required for the storage of local variables for the task and all its nested functions, as well as storage requirements for interrupts (including the nesting).

prio   is the task's priority. Under μC/OS, all tasks *must* have a unique priority number.

OSTaskCreate() returns either OS_PRIO_EXIST if the priority of the task already exists or OS_NO_ERR if the task was created successfully.

A.2.8 OSTimeDly()

OSTimeDly() allows the calling task to suspend itself for an integral number of clock ticks. Valid delays range from 1 to 65,535.

```
void OSTimeDly(UWORD ticks);
```

ticks   specifies the delay in number of clock ticks.

Appendix B

The Companion Disk

The companion disk contains the source code for the larger examples in this book. Enough extra code is included to make the examples functional, and DOS executables are supplied.

B.1 Installation

To install, create an empty directory. The name does not matter, but I use the name \frontp in references to full directory paths. Change directory to the \frontp directory and type

`a:\install`

If you wish to install from the B: drive, type

`b:\install b:`

If you install from a directory on the C: drive, you may enter the drive and directory in which the install script and `files.exe` file have been copied, as in the following example.

`c:\fpdir\install c:\fpdir`

The file called `files.exe` contains a self-extracting archive that is expanded by the install script. The `\frontp` directory should then contain several directories I will describe for you.

Most examples can be ported to another RTOS, compiler, or processor without much difficulty. The exceptions are the examples that use graphics, `multi`, and `fsm`. These use the Borland Graphics Interface (BGI), which may not be available on another system, although you may be able to locate an equivalent drawing library.

Because some of the object files are supplied, they will not necessarily be compatible with the objects you build with your version of the compiler. If you are using a 3.x version of Borland C++, type the command `useb3` from the `\frontp` directory to replace the v4.x object files with files suitable for Borland C++ v3.x. If at some point you upgrade your compiler and want to revert to the v4.x object files, call the `useb4` command from the `\frontp` directory.

Note that if you have a 5.x version of the Borland compiler, it should include Borland C++ v4.5 for use with 16-bit applications. You will need to use v4.5 with these examples.

This installation has been tested with Borland v3.1 and v4.5. If you are using a different version, you may have to change the path the computer searches for include files by editing the makefile for each example.

Although the Borland compiler is required to compile these examples, the built executables are also included so that you can run the examples on any PC. If you wish to recompile any examples, simply call the `make` command from the directory that contains the example you wish to rebuild.

Please report any bugs you find to me at `nmurphy@iol.ie` or via the publisher:

R & D Books
Miller Freeman, Inc.
1601 West 23rd Street, Suite 200
Lawrence, KS 66046
(785) 841-1631

Information on any bugs found will be maintained at `http://www.iol.ie/~nmurphy`.

B.2 The Directory Structure

The examples are divided by chapter number. Three directories are shared among a number of examples — `common`, `include`, and `ucos`. Each example directory contains the source code, a makefile, and the compiled object modules and executable. All code has been compiled with debugging turned on, so you can step through the code with Borland's turbo debugger, if you wish. The examples that use the µC/OS operating system may not be well behaved under the debugger, because the debugger was not designed to be task-aware. Also be aware that if a program using the µC/OS operating system crashes or exits without running the appropriate tidy-up code, the DOS shell will be left in an undetermined state and may crash when you run a subsequent program. For more details about exiting multitasking programs, see Appendix A.

The keys that control the example programs are described on the screen when the program is run. In addition, you can press the Q key at any time to exit any example.

The following sections give a quick overview of how to run each program. You can find a more detailed description of the internals of each example's code in the appropriate chapter.

B.2.1 chap2/tdm

The text display manager (TDM) is a module that allows an application to control the priority and location of strings displayed on a small text display. The TDM structure and the functions that manipulate it are described in Chapter 2.

This example program simulates a two-line by 20-character display on the PC screen. It also displays the full model, showing all three layers of each line. You can use keys to make messages active or inactive. These changes always affect the model, although the 2 × 20 display area will only reflect changes to the highest priority message for each line.

The full model is shown only for illustration purposes. In a real system, the textDisplay() function may be connected to a small LCD text display device.

This example does not use graphics or µC/OS.

B.2.2 chap3/keyq

This example illustrates the queuing of key and other events between tasks. This program displays an imitation of three buttons and three associated LEDs. Each button is associated with a setting. The three settings are temperature, pressure and flow, which control a hypothetical pneumatic process. The settings can be selected with the T, P, and F keys. Once selected, the LED next to the button is illuminated. A numeric display displays the value of the selected setting. The O and L keys imitate arrow keys and increase or decrease, respectively, the currently selected setting.

An alarm area displays warning strings for high temperature and high pressure. For the example, these warnings are put on the queue on a timed basis.

This example uses µC/OS. For this reason, the ucos.lib file in the ucos directory is linked to the executable by the makefile.

B.2.3 chap3/watch

This example illustrates the use of callbacks. Callbacks are associated with the buttons, which change meaning in different modes. The example imitates a digital watch with two modes — a time-of-day mode and a stopwatch mode. The watch has three buttons. The left button changes the mode. The two buttons on the right alter the time of day and start, stop, and reset the stopwatch.

A real digital watch would have a small LCD element to indicate the mode of a symbol. That is not available here, so the mode is simply printed in text below the watch.

This example uses µC/OS. For this reason, the ucos.lib file in the ucos directory is linked to the executable by the makefile.

B.2.4 chap4/fsm

This example illustrates the use of a small finite-state machine (FSM) to control the state of a graphical button. The button was originally designed for a touch screen. For this reason, sliding the mouse onto the button with the left mouse button depressed is the same as depressing the left mouse button while the pointer is over the button. Similarly, sliding off the button releases the button with no change. To change the button from Up to Down, or vice versa, the left mouse button must be released while the pointer is over the button. This allows a user to slide his finger only until the desired button is highlighted. Note that the mouse imitates a finger while the left mouse button is depressed. The mouse has no effect if no mouse button is depressed.

The screen displays a single test button and an exit button. This example requires the BGI to compile. It also links the file egavga.obj from the common directory.

You must have an installed mouse driver to run this program from DOS. A DOS shell spawned from Windows 95 automatically manages the mouse, so this is not a concern.

B.2.5 chap4/menu

This example presents a menu on the PC screen. The hypothetical product is the latest in high-tech footwear. There is a 30-character display along the sole of the shoe, and the owner may configure the trainer using three buttons marked UP, DOWN, and SCROLL. The color of the sole, upper, and lace can be changed at run time, if you pardon the pun. The brightness is set as a number — very useful for joggers who venture out at night. The air pressure in the sole of the trainer is also adjustable via a menu option.

In the menu options, where there are several choices, a double asterisk (**) indicates the currently selected value. As the user scrolls through the possible values, a new one is chosen by pressing the Down (Return) key.

This example does not use graphics or µC/OS.

B.2.6 chap5/multi

This directory contains the `shapes.c` module, which supplies a set of graphical items, including boxes, lines, circles, buttons, and line graphs. The `multi` program is an example that demonstrates some of the capabilities of the `shapes.c` module. The `multi` program allows the user to display two dialogs. The first dialog shows two real-time graphs, which may be frozen using the button below the graph. The second dialog is called Fred. Fred is a match-stick man whose head and body can be moved independently. You can dismember and reassemble him using the keyboard.

This example demonstrates how multidialog graphical user interfaces with buttons and simple animation can be built up from a few basic structures that encapsulate the control of the display.

The BGI library supplies the graphics primitives. It also links the file `egavga.obj` from the `common` directory. This example also uses µC/OS.

You must have an installed mouse driver to run this program from DOS. A DOS shell spawned from Windows 95 automatically manages the mouse, so this will not be a concern.

B.2.7 chap7/multi

This version of `multi` has been recoded in C++. The program's external appearance is identical to the C version used in Chapter 5.

This example uses µC/OS. To use µC/OS with C++, you must make a few changes to allow for the fact that µC/OS is written in C. The calling convention and naming differences between C and C++ are handled by wrapping the include files as follows.

```
extern "C"
{
#include <INCLUDES.H> /* for uC/OS */
}
```

The prototype defined for an interrupt routine is different for C++ than in C. In C, programs use the following form to set an interrupt.

```
setvect(uCOS, (void interrupt (*)(void)) OSCtxSw);
```

In C++ this must be changed to

```
setvect(uCOS, (void interrupt (*)(...)) OSCtxSw);
```

You must have an installed mouse driver to run this program from DOS. A DOS shell spawned from Windows 95 automatically manages the mouse, so this is no a concern.

B.2.8 chap7/tdm

This is the C++ version of the text display manager example presented in Chapter 2. This example does not use graphics or µC/OS.

B.2.9 common

This directory contains the `mouse.c` module that accesses the mouse via the BIOS. This module is linked with the `fsm` and `multi` programs. This directory also contains the `egavga.obj` file, which provides the device driver for the BGI libraries. This file is created from the `egavga.bgi` file using the `bgiobj.exe` program, which is shipped with the Borland compiler.

B.2.10 include

This directory contains header files used by several of the examples. The µC/OS header files are found here. The `mouse.h` file, which allows access to the `mouse.c` module in the `common` directory, is also here. `defs.h` is a header file that contains type definitions that are shared by many examples.

B.2.11 ucos

This directory contains compiled versions of µC/OS. The source code is distributed with the book µ*C/OS: The Real-Time Kernel* (Labrosse, 1992). This version is compiled for the small memory model and can handle up to 63 tasks and five queues. See Appendix A for more information on µC/OS.

B.2.12 borland3

This directory contains the µC/OS operating system in object form, built with Borland v3.1 and a copy of `egavga.obj`, the device driver for the Borland Graphics Interface. You should not have to access this directory directly, but it is used by `useb3.bat`.

B.2.13 borland4

This directory contains the µC/OS operating system in object form, built with Borland v3.1 and a copy of `egavga.obj`, the device driver for the Borland Graphics Interface. You should not have to access this directory directly, but it is used by `useb3.bat`.

B.3 Licensing

You may reuse the source code freely in any nonmilitary application. However, the author and publisher make no warranty of any kind with respect to this software or its fitness for any purpose. If you wish to reproduce any of the source code for publication, please contact R&D Books.

Bibliography

Abrash, Michael. 1996. *Zen of Graphics Programming*, Scottsdale, AZ: The Coriolis Group Inc. ISBN 1-883577-89-6.

Adler, Paul S. and Terry A. Winograd. 1992. *Usability, Turning Technologies into Tools,* New York: Oxford University Press, Inc. ISBN 0-19-507510-2.

Auslander, David M. and Cheng H. Tham. 1990. *Real Time Software for Control,* Englewood Cliffs, NJ: Prentice Hall Inc. ISBN 0-13-762824-2.

Brewster, Stephen, Veli Pekkce Raty, and Atte Kortekangas. 1996. "Earcons as a Method of Providing Navigational Cues in a Menu Hierarchy," *Proceedings of HCI '96,* Springer Verlag London Ltd. ISBN 3-540-76069-5.

Cargill, Tom. 1992. *C++ Programming Style,* Reading, MA: Addison-Wesley. ISBN 0-201-56365-7.

Cargill, Tom. 1994. "Exception Handling: A False Sense Of Security," *C++ Report,* vol. 6, no. 9, November–December.

Connolly, Brian. 1993. "A Process for Preventing Software Hazards," *Hewlett-Packard Journal,* pp. 47–52, June.

Constantine, Larry L. 1995. *Constantine on Peopleware,* Englewood Cliffs, NJ: Yourdan Press. ISBN 0-13-331976-8.

Cooper, Alan. 1995. *About Face: The Essentials of User Interface Design,* Foster City, CA: IDG Books Worldwide Inc. ISBN 1-56884-322-4.

Eckel, Bruce. 1995. *Thinking in C++,* Englewood Cliffs, NJ: Prentice Hall. ISBN 0-13-917709-4.

Ellis, Margaret A. and Bjarne Stroustrup. 1990. *The Annotated C++ Reference Manual*, Reading, MA: Addison-Wesley. ISBN 0-201-51459-1.

European Committee for Standardization. 1995. *pEN794-1 Medical Electrical Equipment — Lung Ventilators*. British Committee for Standardization, 2 Park St., London W1A 2BS.

Foley, James, Andries van Dam, Steven Feiner, and John Hughes. 1996. *Computer Graphics: Principles and Practice, Second Edition in C*, Reading, MA: Addison-Wesley. ISBN 0-201-84840-6.

Gram, Christian and Gilbert Cockton. 1996. *Design Principles for Interactive Software*, London: Chapman & Hall. ISBN 0-412-72470-7.

Gray, Mike. 1992. *Angle of Attack: Harrison Storms and the Race to the Moon*, New York: Penguin Books USA Inc. ISBN 0-14-023280-X.

Hendricksen, C. 1989. "Augmented State Transition Diagrams for Reactive Software, ACM SIGSOFT," *Software Engineering Notes*, vol. 14, no. 6, pp. 61–67, October.

Labrosse, Jean J. 1992. *µC/OS The Real-Time Kernel*, Lawrence, KS: R&D Publications. ISBN 0-87930-444-8

Labrosse, Jean J. 1995. *Embedded Systems Building Blocks*, Lawrence, KS: R&D Publications. ISBN 0-13-359779-2.

Landauer, Thomas K. 1995. *The Trouble with Computers: Usefulness, Usability and Productivity*, Cambridge, MA: MIT Press. ISBN 0-262-12186-7.

Lansdale, M.W. and T.R. Ormerod. 1994. *Understanding Interfaces: A Handbook of Human Computer Dialog*, Cambridge, MA: Academic Press. ISBN: 0-12-52839-0.

Lavery, Dave. 1996. "The Future of Telerobotics," *Robotics World*, Summer. Also available at http://ranier.oact.hq.nasa.gov/telerobotics_page/FutureOfTR.html.

Lethaby, Nick and Ken Black. 1993. "Memory Management Strategies for C++," *Embedded Systems Programming*, July.

Leveson, Nancy. 1995. *Safeware, System Safety and Computers*, Reading, MA: Addison-Wesley Publishing. ISBN: 0-201-11972-2.

Levine, John R., Tony Mason, and Doug Brown. 1992. *Lex & Yacc*, O'Reilly & Associates. ISBN 1565920007.

Lowell, Augustus P. 1992. "The Care and Feeding of Watchdogs," *Embedded Systems Programming*, April.

McGregor, Douglas. 1985. *Human Side of Enterprise*, New York: McGraw-Hill. ISBN 0070450986.

Maguire, Steve. 1993. *Writing Solid Code*, Redmond, WA: Microsoft Press. ISBN 1-55615-551-4.

Meyers, Scott. 1996. *More Effective C++: 35 New Ways to Improve Your Programs and Designs*, Reading, MA: Addison-Wesley. ISBN 0-201-63371-X.

Monk, Andrew. 1995. *Perspectives on HCI: Diverse Approaches,* Cambridge, MA: Academic Press. ISBN 0-12-504575-1.

Musser, David R. and Atul Saini. 1996. *Stl Tutorial & Reference Guide: C++ Programming With the Standard Template Library,* Addison-Wesley. ISBN 0201633981.

Negroponte, Nicholas. 1995. *Being Digital,* London: Hodder and Stoughton. ISBN 0679439544.

Neumann, Peter G. 1995. *Computer Related Risks,* Reading, MA: Addison-Wesley. ISBN 0-201-55805-X.

Norman, Donald A. 1990. *The Design of Everyday Things,* New York: Doubleday. ISBN 0-385-26774-6.

Plauger, P.J. 1996. "State of the Art: Embedded C++," *Embedded Systems Programming,* November.

Preston, Holly Hubbard, and Udo Flohr. 1997. "Global from Day One," *Byte,* vol. 22, no. 3, p. 97, March.

Saks, Dan. 1996. "C++ Theory and Practice: const as a Promise," *C/C++ Users Journal,* vol. 14, no. 11, November.

Saks, Dan. 1997. "C++ Theory and Practice: Placement new," *C/C++ Users Journal,* vol. 15, no. 4, April.

Schmidt, Bobby. 1997. "Let me Say That About this," *C/C++ Users Journal,* vol. 15, no. 5, May.

Schneiderman, Ben. 1993. *Sparks of Innovation in Human Computer Interaction,* Ablex Publishing Corp. ISBN 1-56750-078-1.

Selby, Richard W., Victor R. Basili, and F. Terry Baker. 1987. "Cleanroom Software Development: An Empirical Evaluation," *IEEE Transactions on Software Engineering,* vol. SE-13, no. 9, pp. 1027–1037, September.

Stanton, Neville. 1994. *Human Factors in Alarm Design,* Taylor & Francis Ltd. ISBN 0-74840-0109-1.

Stanton, Neville. 1996. *Human Factors in Nuclear Safety,* Taylor & Francis Ltd. ISBN 07484-0166-0.

Stolper, Steven A. 1995. "Embedded Systems on Mars: The Mars Pathfinder Altitude and Information Management Subsystem," *Proceedings of the Seventh Annual Embedded Systems Conference,* Miller Freeman Inc. ISBN 0-87930-388-3.

Strassberg, Dan. 1994. "User-Interface prototypes help you design products real people can operate," *EDN,* p. 51, March 3.

Stroustrup, Bjarne. 1997. *The C++ Programming Language,* 3rd Edition, Addison-Wesley. ISBN 0-201-41618-2

Thimbleby, Harold. 1990. *User Interface Design,* New York: ACM Press. ISBN 0-201-41618-2.

Vanderheiden, Gregg C. 1996. "Use of audio-haptic interface techniques to allow nonvisual access to touchscreen appliances," Trace R&D Center, University of Wisconsin-Madison, available at `http://trace.wisc.edu/text/kiosks/chi_conf/chi_conf.html`.

Waern, Yvonne. 1989. *Cognitive Aspects of Computer Supported Tasks,* New York: John Wiley & Sons. ISBN 0-471-91141-0.

Wiklund, Michael E. 1994. *Usability in Practice: How Companies Develop User Friendly Products,* Cambridge, MA: Academic Press. ISBN 0-12-751-250-0.

Glossary

clipping   Removing part of a shape or image that is being drawn in order to limit the drawing to a certain area.

custom displays   A front panel that is designed for an application. It contains lights and displays with a particular purpose, distinct from a GUI, which can reconfigure its display at will.

desktop computer   Describes conventional PCs and workstations, which include IBM-compatible PCs, as well as UNIX workstations. Assumed to have a keyboard, mouse, and a windowed GUI.

dialog   Most embedded graphical interfaces do not have windows in the desktop sense. They may, however, have a collection of controls that occupy a predefined area of the display that may appear and disappear at times in reaction to events. Such an independent area is known as a dialog. If the system has several dialogs occupying the whole screen and the user can switch between them, many programmers refer to each dialog as a *screen,* as in configuration screen, settings screen, and so on. I avoid using the word *screen* in this context as it may be confused with the physical screen.

encapsulation   The practice of providing a strict boundary around a piece of data to limit the ways in which it may be manipulated.

filter   Electronically, filters limit the variations of a signal. In software, filtering can be applied by averaging a numerical value to smooth the signal over time. Filtering can also be applied to events to decide which events are passed on to the next level. In some cases, key-down events are passed on for further processing while key-up events are filtered out.

front panel   The physical control area of the device. On many devices, it is actually on the front of the device. In other cases, such as virtual reality headsets or anti-lock brake systems in which the interface is a brake pedal, the physical interface is not available on a panel. However, I use it as the generic term for the physical components of the interface. The front panel may be a graphical touchscreen or simply a row of buttons.

ISR   Acronym for interrupt service routine (ISR). It is a function that is called when the processor is interrupted from its normal flow of control by an external event.

model-view-controller   Describes a paradigm of breaking down the user interface management into three parts. The model represents the application data that the user may manipulate. The view represents the image that the user actually sees. The controller provides the logic to make changes to the model and the view, based on user events.

polymorphism   The ability of an object to be, or to appear to be, different things at different times.

render   To draw. *Draw* can be ambiguous because it may happen at several levels. It may also be restricted by clipping, or the graphic may be drawn to a plane not currently displayed. Rendering refers to the lowest level and means that the pixels are visible to the user. You may draw a circle that is clipped. An arc of the circle is then rendered. This distinction is subtle so, in general, consider *render* to be a synonym for *draw*.

RTOS   Acronym for real-time operating system, a multitasking scheduler suitable for systems with real-time constraints.

soft keys   Mechanical keys placed next to a text or graphics display in such a way that the software can change the meaning associated with the key by changing the text displayed next to it. Usually, there is no text or icon printed on the key itself.

task   In a real-time operating system, this generally means a thread of control. Several tasks may be active simultaneously, and the RTOS arbitrates the CPU time among them. Tasks commonly live in the same address space and can access the same data. They can communicate via queues, semaphores, and other RTOS-supported mechanisms.

Index

A
abbreviations 151
Abrash, Michael 82
access function 161
ADA 287
ADC, see analog-to-digital converter
Adler, Paul S. 235
Air Safety Reporting System 279
alarm 252
    audible 274
    fail silently 277
    false 276
    standing 276
analog-to-digital converter 165
animation 96, 108, 130
Apollo 11 138
`Area` 87
Armstrong, Neil 138
assembler 8, 127
`assert` 138, 175
AT&T 138
Atari 800 127
ATM, see automatic teller machine
automatic teller machine 242, 245, 249, 252, 254, 260, 265, 268
automation, home 285

B
BGI, see Borland Graphics Interface
BIOS 82, 302
bitmaps 83
Borland Graphics Interface 99, 298
Brewster, Stephen 68
bubble help 249
BYTE magazine 154

C
calculator 256
callbacks 126
camera 240
Cargill, Tom 158, 187
`catch` clause 172
chemical reactions 220
chess 220
cleanroom methodology 217
clipping 84, 85
code reuse 82, 127, 179, 180, 186, 199, 202, 215
coding conventions 4
collision avoidance system 285
collision detection 108
color blindness 263
command line 132, 237
commensurate effort 246
Commodore 64 127
compiler 152, 157
    optimization 197
conceptual model 260
confirmation 269
    of actions 245, 271
    of notification 264
Connolly, Brian 141
`const` 60, 152, 194
context 83
context switch 30, 289
control loop 150

311

Cooper, Alan 260
coordinate system 84, 93
cost 81, 144, 154, 155, 157, 176, 188
cowboy programmers 127
"creeping featurism" 202
cruise control 256
cry wolf 272, 276
cursor location 122

D

dashboard 227, 233, 254, 257
dead man control 270
dead space, between buttons 123
debouncing 27
`delete` operator 164
`delete[]` 197
desktop computers 1
dialog 82, 211, 264, 271
digital watch 49
direct manipulation 262
dirty flag 97
Disneyland 240
distance viewing 131
domain expert 240
`Drawable` 87
drawing tool 58
duty cycle 262

E

earcon 68
early binding 170
ease of use 221, 233, 237, 245, 246
Eastman Kodak 240
Eckel, Bruce 158
Ellis, Margaret A. 158
ethnography 233
event loop 126, 209
event-driven 2
exclusive-or 120
eyes-off-road time 233

F

false alarms 151
feedback 146
    initial 130
    loop 224
fidelity 229
field trial 232
filtering 150, 257, 276
    input event filters 56
flicker 85, 117
Flohr, Udo 154
fly by wire 257
Foley, James 82
fonts 83, 153
    scalable 83
function pointers 50
furnace 142, 149, 255

G

`getch` 37
global positioning system 236
GPS, see global positioning system
Gram, Christian 26
graphics 283
graphics context 83
Gray, Mike 138

H

hard real time 56, 184
Hawthorne effect 240
HD44780 18
heap 90, 158, 160, 164, 179, 181, 184, 198
    fragmentation 23, 90
heel rest 268
Hendricksen, C. 57
hierarchy 94
high-risk mode 246
Hitachi HD44780 154
home automation 285

Honda 234
hot-spot 122
HTML 134

I

icon 122, 155, 248
information kiosk 268
infusion pumps 136
International Standards Organization 282
Internet 229
interoperability 286
interrupts, disabling 32
ISO 8601 251
ISO 9995 251
ISO, see International Standards Organization
ISR 30, 292

J

Java 287
jitter 123

K

kbhit 37

L

labeling 153, 155, 224, 231, 248, 249, 250, 260, 265, 274, 275
    overlay 237
labels 153, 155
Labrosse, Jean J. 14, 18, 27, 154, 289, 291, 302
Landauer, Thomas K. 244
late binding 170
Lavery, Dave 285
LCD 117, 257, 258, 264
Lethaby, Nick 186

Leveson, Nancy 137, 139, 141, 144, 269, 279
Levine, John R. 134
Lex 134
load balancing 131
localization, see translation
locks 40
Lowell, Augustus P. 145
lung ventilator 131, 241, 270

M

macro 9
    assert 138
    GET_DRAWABLE 89, 212
    versus in-line function 179
makefile 153
malloc 60, 91
map 8, 13, 17, 19, 23, 27, 34, 83, 87, 123, 197, 220, 236, 251, 267, 284
Mars Pathfinder spacecraft 145, 287
Maxim ICM7218 139
McGregor, Douglas 234
Mealy FSM 57
memory management 23, 181
memory pool 185, 208
mental model 238
message passing 54
Meyers, Scott 158
µC/OS 289
microwave oven 248, 256, 264
Monk, Andrew 233
Moore FSM 57
motion device 30, 32, 33, 44, 189
mouse 27, 44, 231, 249, 258, 260, 265, 268, 302
    drag-by-outline 120
    dragging 111, 113, 123
    versus pen 260, 267
    versus trackball 231, 260
MS-Windows 3, 130
multi 86

multiple inheritance 191
multiplexing 147, 250, 254, 256
    LEDs 43
Musser, David R. 188

N

name collisions 178
namespace 178, 224
needle gauge 258
Negroponte, Nicholas 282
Neumann, Peter G. 287
`new` operator 164, 177, 181, 195, 198, 208
`new[]` 182
next-bench syndrome 240
nuclear power plant 230

O

opaque type 21, 89
optimization 18, 103, 105, 179, 186
origin 84
Orwell, George 224
oscilloscope 116
overlapping objects 96

P

paper-cutting machine 270
parallax 260, 265
PC clock 118
pen 260
photocopier 240, 250
pilot 137, 138, 149, 227, 233, 235, 268, 271, 278, 285
placement `delete` 187
placement `new` 195
Plauger, P.J. 188
pneumatic system 227
polling 145
polymorphism 89

Preston, Holly Hubbard 154
private data 160
psychology 67, 240
public data 160
pure virtual 189

Q

QWERTY 251

R

read ahead 44
read-only memory 152
real estate 249
real-time graph 116
reboot 134, 187
recursion 211, 237
redundancy 149, 269, 278
    two monitoring thermistors 272
resource 40, 185, 269, 290
reverse engineering 25
ROM 194, 195
root `Container` 124, 210
root container 93

S

Saks, Dan 195, 197
scale 84
Schmidt, Bobby 161
Schneiderman, Ben 253, 268
scrolling 117, 130
Selby, Richard W. 217
self-test 277
service engineer 153, 237
Sinclair ZX Spectrum 245
slave processor 130
Smalltalk 12, 15, 54
soft keys 155, 265
software failures 137
sound 68

speedlist 268
stack stereo 252
Stanton, Neville 230, 276
`static` 10
static electricity 148
Stolper, Steven A. 145
Strassberg, Dan 229
Stroustrup, Bjarne 158
surface area 250

T

talking finger 268
tape deck 238, 246, 247, 254
technology-driven 244
telephone 251
telerobotics 285
testing 136
text display manager 19
Therac-25 144, 269
thermistor 146
Thimbleby, Harold 220, 246, 256
`this` 201
thread safe 118
throw 172
timeline 268
time-out 271
tolerance 239, 258
touch-sensitive areas 122
Tower of Babel 283
trackball 259, 260
translation 151, 248
type ahead 29, 33

U

undelete 245
undo 58, 246
`union` 126
usability lab 232
usability professionals 232
`USING_BGI` 99

V

van Dam, Andries 82
Vanderheiden, Gregg C. 268
VCR 251, 256, 259, 264
ventilator, see lung ventilator
verbal commands 284
version control 2
VGA 40, 81, 82
video recorder 249
voice
    control 284
    mail 250
    synthesis 284
`volatile` 197
voting mechanisms 149

W

Waern, Yvonne 67
walk-up application 267
Web-enabled 283
Wiklund, Michael E. 241
wind shear detection 285
window manager 264
Wizard of Oz 243

X

X-bitmap 83
`.xbm` 83
Xerox 240
X-Window 3, 127

Y

Yacc 134

Expert Advice When You Need It Most!

The File Formats Handbook
by Günter Born

More File Formats for Popular PC Software
by Jeff Walden

Graphics File Formats, 2nd Edition
by David C. Kay & John R. Levine

Internet File Formats
by Tim Kientzle

Inside Windows File Formats
by Tom Swan

File Formats
by Allen G. Taylor

Essential Books on File Formats CD-ROM

Look no further! The *Essential Books on File Formats* CD-ROM contains the complete text from six books, which will provide you with the most comprehensive and detailed information on all the important and popular file formats in use today. Selected by the editors of *Dr. Dobb's Journal,* this CD-ROM contains invaluable information on file formats used for graphics, multimedia, sound, databases, spreadsheets, Windows, the Internet, and much more!

No matter what your programming focus, the *Essential Books on File Formats* CD-ROM discloses the secrets and insider knowledge you need to make your programming instantly compatible with all the major applications right from the start. Plus, the CD-ROM's powerful full-text search engine and hyperlink capabilities allow you to search quickly and easily across all the books to link directly to the information you need.

Price: $69.95

To Order Call: 800-500-6739
U.S. & Canada

All Other Countries: 913-841-1631
E-mail: orders@mfi.com
Fax Orders: 913-841-2624

Mail Orders:
Dr. Dobb's CD-ROM Library
1601 W. 23rd St., Ste. 200
Lawrence, KS 66046-2703 USA